exploration in future food-processing techniques

CONTRIBUTORS

Leo Friedman
Samuel A. Goldblith
Marcus Karel
Joseph J. Licciardello
John T. R. Nickerson
Gerald Silverman
Emily L. Wick

EDITOR

Samuel A. Goldblith

The M.I.T. Press, Massachusetts Institute of Technology, Cambridge, Massachusetts, 1963

Library of Congress Catalog Card Number: 63-20865
Printed in the United States of America

PREFACE

This symposium was planned early in 1963 and held on May 2 to present some of the research progress to the segments of industry and of government that have been generous in sponsoring research in food science and technology at the Massachusetts Institute of Technology.

After the program had been planned and the preparation of the papers well under way, the editor felt that there might be some value in publishing these papers as a monograph in order that they might become available to broader segments of the food industry.

The cooperation of the Industrial Liaison Office at M. I. T., particularly Mr. Lloyd S. Beckett, Jr., and Mr. Vincent A. Fulmer, its Director, who has been on leave as Executive Assistant to the Chairman of the M. I. T. Corporation, is gratefully acknowledged. The editor also wishes to thank the participants, who made this program possible, and Dr. Nevin S. Scrimshaw, the Head of the Department of Nutrition and Food Science, who had much to do with its initial planning and, in fact, made the suggestion for holding such a symposium in food science and technology.

To Miss Janet Anderson the editor expresses his appreciation for her assistance in preparing this manuscript and to Miss Ann Peterson for her capable editorial assistance in the preparation of the original papers.

Cambridge, Massachusetts
May 6, 1963

Samuel A. Goldblith

CONTENTS

Chapter 1

INTRODUCTION

Samuel A. Goldblith

It is fitting that this symposium be sponsored by the Industrial Liaison Office at M. I. T., for this Institute has long recognized the unity and oneness of science and technology. It has long recognized the role of industry and industrial technology in advancing science itself.

Today, M. I. T. has the Industrial Liaison Program to assist in the creation of an intermingling, symbiotic, and synergistic relationship between the Institute and American industry. This has proved, I feel, to be mutually beneficial. Although this program has been in existence for only a decade, its spirit dates back many decades. In the field of food science and technology, then called "industrial biology," it dates back to the 1890's, when canning was placed on a scientific basis through the pioneering researches of Samuel C. Prescott and William Lyman Underwood at a place known familiarly to many as "Boston Tech." To this audience these two names and their work need no explanation.

Dr. A. W. Bitting, in his classic book Appertizing,* has said that "the canners of this country owe more to Prescott and Underwood for placing their industry upon a scientific basis than to any other investigators ... their first three papers were epoch-making in changing an industry based upon individual experience to one under scientific control."

The food and container industries recognized this pioneering work of an industrialist and a university professor. One of our Industrial Liaison Program members, the American Can Company, established the first laboratory for thermal processing research in 1906; the canning industry itself followed with the establishment of the National Canners' Association Laboratory in 1913, and I might add with justifiable, and I hope forgivable, pride that one of its outstanding research directors for many years, Dr. E. J. Cameron, was an alumnus of this Institute and a student of Dean Prescott.

Prescott's and Underwood's pioneering researches in canning were followed by many investigations here in freezing-preservation

*The Trade Pressroom, San Francisco, California, 1937, p. 50.

of foods and in dehydration by Prescott and later by Dr. Bernard E. Proctor and his students.

Professor Proctor and his students in the 1940's and 1950's pioneered at M.I.T. in irradiation preservation of foods, a large proportion of the funds for these studies coming from the food and container industries and through the Industrial Liaison Program. Today, we begin to see irradiation preservation of foods emerging as still another and possibly important method of preservation. The recent publication of a regulation permitting the use of ionizing energy to sterilize bacon is a historic milestone. The decision legalizing the manufacture and sale of this preserved product, based solely upon proven scientific facts, now has rekindled interest among members of the food industry in this fascinating field of preservation, which has been one of the chief research interests of many of us for almost twenty years.

In this symposium we are presenting some of the existing research programs in the radiation preservation of foods and in freeze dehydration. By no means are these inclusive. They do not represent the entire research program in food science and technology here at M.I.T., nor does time permit the presence of all of our research staff on this program.

These papers, however, will illustrate some of the integrated research programs now active in the department on these two intriguing methods of food preservation. May I take this opportunity to reiterate our thanks to the various governmental and industrial organizations for their generous support of the research discussed here.

In most food preservation methods, science has followed technology — the university has followed industry. Radiation preservation of foods, however, is a notable exception to this generalization.

As many are aware, one of the key problems which has long stymied the utilization of ionizing energy for preservation of foods is the undesirable secondary side reaction induced by free radical and activated molecule formation in foodstuffs as evidenced by changes in color, texture, and flavor.

Many approaches have been suggested, among which is that of concurrent radiation-distillation. We, in this department, have been concerned with this method for almost ten years.

Ultimately, the solution of this problem will depend upon the basic research that will unlock some of the elusive secrets which are still within the molecules and are related to flavor, color, and texture. Such a basic program is under way here; for the past six years, this program has been spearheaded by Dr. Emily L. Wick, an organic chemist, whose research field is the chemistry of flavor Dr. Wick's paper on the "Volatile Components of Irradiated Beef" discusses the concurrent radiation-distillation technique and reviews the knowledge of the radiation-induced chemical changes in beef.

One of the methods available to reduce the undesirable side reactions induced by ionizing energy is that of simply reducing the dose to which the food is exposed. To achieve this, we can employ one of perhaps three or four techniques. One of these, the use of bacterial sensitizing compounds, has been studied by Dr. Gerald Silverman and his associates here at M.I.T. for the past three years.

Dr. Silverman's paper discusses radiosensitizers, reviews this field, and presents some interesting data on vitamin K_5 and its derivatives.

Still another method of possibly reducing the undesirable side reactions induced when ionizing energy bombards foodstuffs is the use of the complementary effects of thermal energy and ionizing energy. Some time ago, work by Morgan, Reed, Kempe, Kan, and others indicated that this technique is one which allows for utilization of lower doses of ionizing energy by using a relatively small amount of thermal energy (sublethal quantities), or to put it another way, the utilization of sublethal doses of ionizing energy will permit the utilization of relatively small amounts of thermal energy.

The practical implications of this technique are obvious in the meat industry where thermal sterilization of large cans of meat products is impossible because of the difficulty in putting sufficient quantities of energy into the center of the can without overheating the outermost layers. Dr. Licciardello's paper discusses work on the complementary effects of thermal and ionizing energy and also presents some interesting data on the advantage of simultaneously applying both types of energy using an organism of important present-day public health implication, Salmonella typhimurium.

The paper by Professor Nickerson describes and exemplifies the unity of science and technology. It illustrates, too, the potentialities of ionizing energy as a means of food preservation (in contrast to sterilization). The ocean contains a vast source of food supply which as yet has not been fully exploited. Professor Nickerson's paper describes a research program which may, indeed, open up a means of greater utilization of sea food products in this country.

Among the newer methods of food preservation, freeze dehydration of foods has received a great deal of publicity and has had an enthusiastic response by the food industry and the consuming public.

While this method is in industrial use today by the food industry, there are a number of considerations relative to these products, their manufacture, and their reaction with moisture and oxygen which need careful attention. Dr. Karel's paper discusses the various unit operations involved in freeze dehydration and their effect upon the quality of the final lyophilized foodstuff. The discussion is illustrated with experimental data.

The paper by Dr. Goldblith considers some of the microbiological aspects of freeze dehydration and delves into many of the unsolved problems as well as pointing out the tremendous quantities of microbiological data which have been obtained in the freeze-dehydration industry.

The Food Additives Amendment of 1958 awakened many people to a realization of the need for formal recognition of "food toxicology," "wholesomeness," or "food safety." Recognition of the field by universities again lagged behind industrial recognition. This is not surprising, inasmuch as the work done in the field of food safety had been done by the food and container industries using techniques developed largely by them and the drug industry in conjunction with the Food and Drug Administration. These techniques are costly procedures, by and large, and have been the subject of so much discussion that a review of this subject at the present time would prove to be superfluous and redundant. There are, however, some basic points which should be considered and which are relevant:

1. The test procedures now in use have been developed largely by toxicologists and pharmacologists for use with drugs, and they have been adapted for foodstuffs and food additives.
2. These tests are costly and of long duration.
3. Our laws, including those dealing with foods, are usually written by politicians (and here I refer to this word in its highest sense), and not by scientists.
4. Thus, if we do not like our present laws, we should do something other than telling the Food and Drug Administration. This agency simply does not make the laws.

What can universities do about the points which I have raised? The Massachusetts Institute of Technology has recognized the problems and the obligations of the university. It has formally established within the Department of Nutrition and Food Science a group in food safety under the leadership of Dr. Leo Friedman, whose paper presents some of the problems in the wholesomeness evaluation of foods and outlines the research program in this field and its rationale.

Chapter 2

VOLATILE COMPONENTS OF IRRADIATED BEEF*

Emily L. Wick

Radiation preservation has been a subject of interest in recent years because of its ability to inactivate and destroy microorganisms, and its consequent potential contribution to the solution of food distribution problems by permitting extended storage at nonrefrigerated temperatures. Radiation may, however, change the flavor, appearance, texture, and therefore the acceptability of a food. Its exact effect depends upon the food in question, the processing conditions, and the total radiation dose employed. Preservation of beef by irradiation is of special interest since beef is particularly sensitive to the development of undesirable flavor and odor. The chemistry of certain of these changes is the subject of this paper. Details of the research were presented in September 1962 (Mizutani and Wick, 1962) before the American Chemical Society and are being prepared for publication.

Any basic study of the chemistry of these changes is complicated by the fact that knowledge of the flavor components of nonirradiated beef is also required. Otherwise, a determination of which components are produced by irradiation (and thus may be responsible for off-flavor) and which components are normally present in nonirradiated beef cannot be made. Knowledge of both is required if the mechanism of off-flavor formation is to be understood.

Existing knowledge of the volatile components of beef-irradiation odor as produced in raw and in enzyme-inactivated lean beef has been accumulated by several research groups.

The first few tables will summarize the constituents of irradiated raw beef which have been identified. These compounds were isolated from ground lean beef samples which exhibited irradiation off-odor as a result of doses of from 2 to 6 megarads. Methods of isolation were either gas entrainment of volatiles from aqueous slurries, vacuum distillation, or concurrent radiation-distillation. Their identification is the result of research

---

*The work herein has been done under Contract No. DA-19-129-QM-1374, administered by the Natick Laboratories, Department of the Army, Natick, Massachusetts.

by Burks et al. (1959), Merritt et al. (1959), Stahl (1957), the group at the American Meat Institute Foundation (Batzer et al., 1955), and our own group at M.I.T. (Wick et al., 1961a). Table 2.1 lists the basic and alcoholic constituents of raw irradiated beef.

Table 2.1. Volatile Components of Irradiated Raw Ground Beef (Basic and Alcoholic Constituents)

Ammonia*	Pyrrole
Methyl amine	Pyridine
Ethyl amine	Aniline
4 Unknown amines	
Methanol*	
Ethanol*	

*Also present in nonirradiated beef odor isolates but in smaller quantity.

Of these compounds, ammonia, one of the unknown amines, methanol, and ethanol are components of nonirradiated raw beef. Table 2.2 shows sulfur-containing components of irradiation off-flavor odor isolates. Hydrogen sulfide, methyl and ethyl

Table 2.2. Volatile Components of Irradiated Raw Ground Beef (Sulfur-Containing Components)

Hydrogen sulfide*	Methyl ethyl sulfide
Methyl mercaptan*	Methyl isopropyl sulfide
Ethyl mercaptan*	Diisopropyl sulfide
n-Propyl mercaptan	Dimethyl disulfide
Isobutyl mercaptan	Diethyl disulfide
A C_5 mercaptan	Ethyl isopropyl disulfide
3-(Methylthio)-propionaldehyde	Diisopropyl disulfide
Dimethyl sulfide*	Carbonyl sulfide
Carbon disulfide	

*Also present in nonirradiated beef isolates but in much smaller quantity.

mercaptans, dimethyl sulfide, and dimethyl disulfide are also found in nonirradiated beef but in much smaller quantity. Carbonyl components are given in Table 2.3. Of these, acetaldehyde, propanone, and 2-butanone are present in nonirradiated beef in smaller amounts. A number of hydrocarbons (listed in Table 2.4) have been identified as components of irradiated raw beef. It has been generally observed that the yield of volatile components increases with increased doses of radiation.

The relative contribution to irradiation off-odor of individual members of this, no doubt, incomplete list of compounds is unknown. It has been believed that the sulfur- and nitrogen-containing substances must contribute significantly because of their natural strong, unpleasant odors. Merritt (1961) has

Table 2.3. Volatile Components of Irradiated Raw Ground Beef (Carbonyl Components)

Acetaldehyde*	Pentenals
2-Propenal	Hexenals
2-Butenal	Propanone*
Pentanals	2-Butanone*
Hexanals	3-Buten-2-one
Heptanals	

*Also present in nonirradiated beef isolates, but in smaller quantity.

Table 2.4. Volatile Components of Irradiated Raw Ground Beef (Hydrocarbons)

Ethylene	*n*-Heptane
Propene	*n*-Octane
1-Butene	Benzene
1-Pentene	Toluene?
1-Hexene	Ethylbenzene?
1-Heptene	Isopropylbenzene?
1-Octene	Trimethylbenzene?
Propane	Butylbenzene?
n-Butane	
n-Pentane	
n-Hexane	

suggested that dimethyl sulfide, 1-hexene, and *n*-hexane are important components of irradiation odor and has pointed out that the quantity of these compounds produced increases directly with the radiation dose.

It is of interest at this point to digress from the constituents of irradiation odor to a very brief review of the presently known volatile constituents of beef flavor as it is recognized in beef extract or beef broth. The substances listed in Table 2.5 have been identified in various aqueous extracts of cooked ground beef. The fact that most of these components of one of the most acceptable odors known are also found in the bland aroma of raw beef, and in the unpleasant odor of irradiated beef, emphasizes an important lesson now well known to flavor chemists. We have learned, thanks to modern instrumentation, that complex mixtures of volatile components can be detected quite easily in almost every foodstuff. We have also learned that the components of these mixtures can be identified. Many of the same components are found in a wide variety of foodstuffs. Our major challenge is, therefore, to select from the many components present those which contribute to the odor or flavor under investigation. This will no doubt require careful quantitative as well as qualitative analysis and, in addition, characterization of odor and flavor precursors.

Table 2.5. Volatile Components of Beef Flavor (Cooked Ground Beef)

Ammonia[1,3]	2-Methylpropanal[2]
Methylamine[3]	3-Methylbutanal[2]
Hydrogen sulfide[1,2,3]	Propanone[1,2,3]
Methyl mercaptan[2]	2-Butanone[2]
Ethyl mercaptan[2]	Diacetyl[1]
Dimethyl sulfide[1,2]	Formic acid[1]
Formaldehyde[3]	Acetic acid[1]
Acetaldehyde[1,2,3]	Propanoic acid[1]
Propanal[2]	2-Methyl propanoic acid[1]
	Lactic acid (NH_4 salt)[3]
	Methanol[2]
	Ethanol[1]

1. M. H. Yueh and F. M. Strong, J. Agr. Food Chem. 8, 491 (1960).
2. A. E. Bender, Chem. and Ind. 52, 2114 (1961).
3. I. Hornstein et al., J. Agr. Food Chem. 8, 65 (1960).

Research on the precursors of beef odor and of irradiation off-odor has been carried out at the Meat Institute Foundation (Batzer et al., 1961). The investigators there have fractionated beef and subjected the resulting fractions to heat or irradiation in order to determine which contained precursors to the odors in question. Their fractionation scheme is outlined in Table 2.6. Diffusate A from the dialyzable water-soluble extractives of lean ground beef yielded on further separation on Sephadex G-25 and then on Dowex-1 columns a fraction which exhibited, when boiled, a beef broth odor and flavor. When heated in fat, a broiled steak odor was produced. The major flavor precursor present in this fraction is believed to be a "glycoprotein," which yields on hydrolysis glucose, a mixture of amino acids, inosinic acid or inosine, and phosphoric acid. The fact that none of these constituents contained sulfur is of interest since it indicates that sulfur-containing volatile components of beef odor isolates may not be fundamental contributors to beef odor!

Precursors to irradiation odor were found in residue E when either raw beef or beef which has been cooked to an internal temperature of 160°F, was fractionated. Further fractionation of E yielded an unknown white solid, probably a phospholipid, which strongly exhibited irradiation odor in aqueous solution. It contained nitrogen and a carbohydrate but no sulfur. Again the contribution of sulfur compounds to the off-odor is challenged. Fraction B_4 from irradiated cooked beef also contained precursors to a weak irradiation odor. Its composition was not studied.

Analysis of each fraction after irradiation, for hydrogen sulfide, methyl mercaptan, acid-soluble carbonyl compounds, and

Table 2.6. Fractionation of Beef-Flavor Precursors*

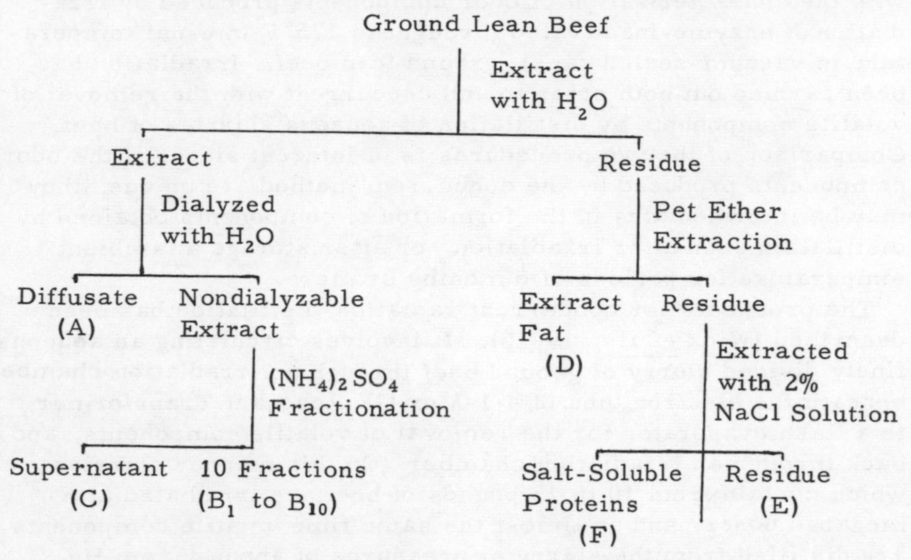

*O. F. Batzer, A. T. Santoro, and W. A. Landmann, J. Agr. Food Chem. 10, 94 (1962).

nonvolatile sulfhydryl compounds showed that fractions B_6 and B_7 from raw meat and fraction A from cooked meat contained precursors to hydrogen sulfide and mercaptan. The main source of these compounds in irradiated raw beef appeared to be proteins precipitated in the range of 50 to 80 per cent saturation with ammonium sulfate. Cooking apparently altered these proteins so that no sulfur compounds could be evolved. Nonvolatile sulfhydryl compounds were detected in fractions A and C in fresh and cooked beef and, in addition, in F in cooked meat. Only fraction A from cooked meat evolved both hydrogen sulfide and methyl mercaptan concurrently. On the basis of these observations, it was concluded that nonvolatile sulfhydryl components (such as glutathione) were not the major precursors of volatile sulfur compounds. This was also confirmed by work with radioactive compounds (Martin et al., 1962).

Carbonyl compounds were evolved in greatest quantity from fractions A and C from fresh meat, and from B_7, B_8, and C from cooked beef. Only very small quantities were detected on irradiation of the lipid fractions.

On the basis of this very brief review of previous knowledge of the components of beef-irradiation odor and of their precursors, it is evident that much remains to be learned. It is also clear that if investigations of odor precursors and odor components can be carried on concurrently, progress in this field can be accelerated.

Recent investigations in our laboratories have been concerned with the characterization of odor components produced by irradiation of enzyme-inactivated (brought to 175°F internal temperature in vacuum-sealed cans) ground lean beef. Irradiation has been carried out both prior to and concurrent with the removal of volatile components by distillation of aqueous slurries of beef. Comparison of the two procedures is of interest since, if the odor components produced by the concurrent method are unique, they may be intermediates in the formation of components obtained by distillation soon after irradiation, or after storage at ambient temperature for periods of 6 months or more.

The procedure for concurrent radiation-distillation has been described (Wick et al., 1961b). It involves circulating an aqueous finely divided slurry of ground beef through an irradiation chamber beneath the electron tube of a 1-Mev GE Resonant Transformer, to a flash evaporator for the removal of volatile components, and back through an irradiation chamber. In this manner slurries which contain from 10 to 15 pounds of beef are irradiated at 5-megarad doses, and at almost the same time volatile components are distilled from the slurry at pressures of about 25 mm Hg. The average temperature of the circulating mixture is 32-36°C, and the rate of distillation is approximately 6 liters an hour. When irradiation has been carried out prior to distillation, cans of cooked beef were subjected to 5 megarads in the M.I.T. cobalt-60 source. They were then opened and the beef slurried and distilled in the manner just described.

The volatile substances were collected in an ice bath which condensed a major portion (8 liters) of the distillate, in a dry ice-ethanol bath which trapped about 50 ml of condensate, and in liquid nitrogen traps which collected only vapors. In the case of concurrent radiation-distillation an additional condensate — called distillate I — was collected while stable distillation conditions were being attained and before irradiation was begun. Since distillate I contained only components distilled from nonirradiated beef, it served as a kind of "internal standard" for these preparations.

The contents of the liquid nitrogen traps in all irradiated beef preparations exhibited strong mercaptan-like odors. Gas chromatography of the vapors showed the presence of many very volatile components. Treatment with mercuric chloride solution caused precipitate formation. Further investigation of these substances was not carried out.

A measure of the quantity of oxidizable organic substances present in the aqueous beef distillates was obtained by the method of Gertner and Ivecovic (1954). Representative results of this determination and the yields of ether-extractable components of the distillate are summarized in Tables 2.7 and 2.8.

The total quantity of oxidizable material found in the distillates and traps from the concurrent process (Table 2.7) was essentially

Table 2.7. Comparison of Distillates and Ether-Soluble Odor Isolates Obtained by Concurrent Radiation-Distillation

Preparation	Weight of Beef (lb)	Distillate	CO_2/lb Meat (mg)	Odor Isolate/lb (mg)	Total Weight Odor Isolate
1. Nonirradiated	11	I	12.5	7.8	85.8
		II	32.0	6.5	71.9
		Traps	9.5	2.2	24.1
			Total 54.0	Total 16.5	Total 181.8
2. Irradiated	12	I	19.0	5.4	64.3
		II	27.3	5.3	63.3
		Traps	7.4	4.7	56.3
			Total 53.7	Total 15.4	Total 183.9
3. Water Blank	0	Total Condensate	2.5*	——	7.7

*Calculated as though 14 pounds of beef had been processed.

the same (about 54 mg CO_2 per pound of beef) from both the irradiated and nonirradiated preparations. Similarly, the total yield of ether-extractable odor isolates was approximately the same (about 16 mg per pound of beef) in both cases. The only obvious difference between the preparations was their odor. The contents of distillate II and of the traps from irradiated preparations always strongly exhibited irradiation off-odor. The same fractions from the nonirradiated preparation and distillate I from both preparations had bland, meatlike odors.

A blank preparation was used in which only water was processed. Oxidizable carbon was found in this condensate. It and its ether extract possessed no odor. The extract showed the presence of two trace substances when gas-chromatographed, indicating that the organic material present was essentially ether-insoluble. The

Table 2.8. Comparison of Distillates and Ether-Soluble Odor Isolates Obtained by Radiation Prior to Distillation

Preparation	Weight of Beef (lb)	Condensate	CO_2/lb Meat (mg)	Odor Isolate/lb (mg)	Total Weight Odor Isolate
1. Nonirradiated	15	Distillate	10.0	5.4	80.9
		Traps	5.0	5.9	89.9
			Total 15.0	Total 11.3	Total 170.8
2. Irradiated	15	Distillate	11.2	6.7	100.5
		Traps	6.3	6.9	103.4
			Total 17.5	Total 13.6	Total 203.9
3. Irradiated	9	Distillate	45.3	18.2	163.9
		Traps	8.9	5.5	49.2
			Total 54.2	Total 23.7	Total 213.1
4. Water Blank	0		——	——	30.0

presence of organic material as an artifact is thus not believed to be cause for concern.

A summary of data concerning preparations that employed irradiation followed by distillation (that is, a nonconcurrent radiation-distillation procedure) is given in Table 2.8. The total yield of ether-extractable odor isolates was roughly 11 to 23 mg per pound of beef in both nonirradiated and irradiated preparations. This was comparable (within experimental error) to the yields obtained by the concurrent procedure. Analysis of the ether extract of a water blank showed that, though it possessed no irradiation odor, it contained several components that might occur as artifacts in beef odor isolates.

Separation of odor isolates has been carried out by isothermal and by temperature-programmed gas chromatography. The resulting fractions were trapped in U-tubes immersed in liquid nitrogen, and then vacuum-transferred to capillary tubes which were sealed and stored under refrigeration for later analysis. The minimum number of components present in each fraction was determined by rechromatography of the vapor over the fraction on a 5% Ucon 50HB-2000 column attached to an ionization detector. The infrared spectrum of each fraction was obtained if it could be isolated in sufficient amount. The resulting infrared sample was then rechromatographed to gain further knowledge of the component or components present.

Figure 2.1 shows typical chromatograms obtained by temperature-programmed separation of isolates from the distillate and traps of

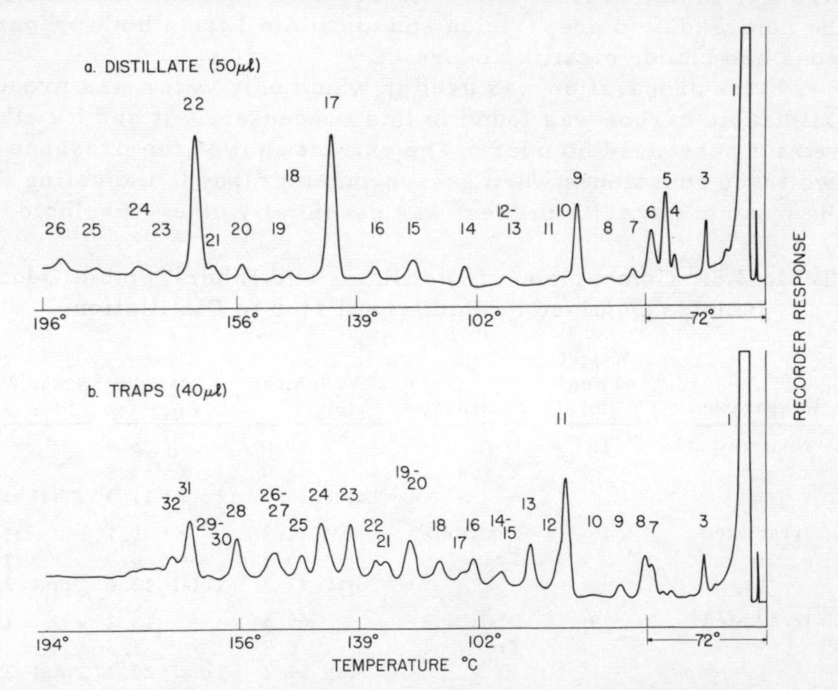

Figure 2.1. Temperature-programmed separation of irradiation flavor isolates on a 20% Carbowax 20M column

beef which had been irradiated at the Co-60 source and then distilled. As you can see, the distribution of volatiles is quite different in each isolate. This is a fortunate circumstance since it simplified the separation and isolation of individual components. Based on the identity of their infrared spectra and retention data with those of authentic reference compounds, the following definite identifications have been made. In the distillate:

Fraction	Identity
5	2-Butanone
7	Benzene
9	2-Butanol
12-13	*n*-Hexanal
14	*n*-Butanol
15	*n*-Heptanal
16	*n*-Pentanol
17	Acetoin and *n*-octanal
20	*n*-Nonanal
22	Methional (3-(methylthio)-propionaldehyde)
24	Benzaldehyde

The remaining fractions are as yet unknown. In the traps:

Fraction	Identity
7	*n*-Nonane
9	1-Nonene
16	*n*-Undecane
18	1-Undecene
19-20	*n*-Heptanal
28	*n*-Nonanal

Several fractions were only tentatively identified since their spectra were not completely in agreement with reference spectra even though their retention data were satisfactory. Thus tentative identification was assigned to the following fractions. In the distillate:

Fraction	Identity
6	Ethanol

In the traps:

Fraction	Identity
12	*n*-Decane
21	*n*-Dodecane
23	1-Dodecene

An interesting comparison is made in Figure 2.2 between chromatograms of the distillate isolates from nonirradiated

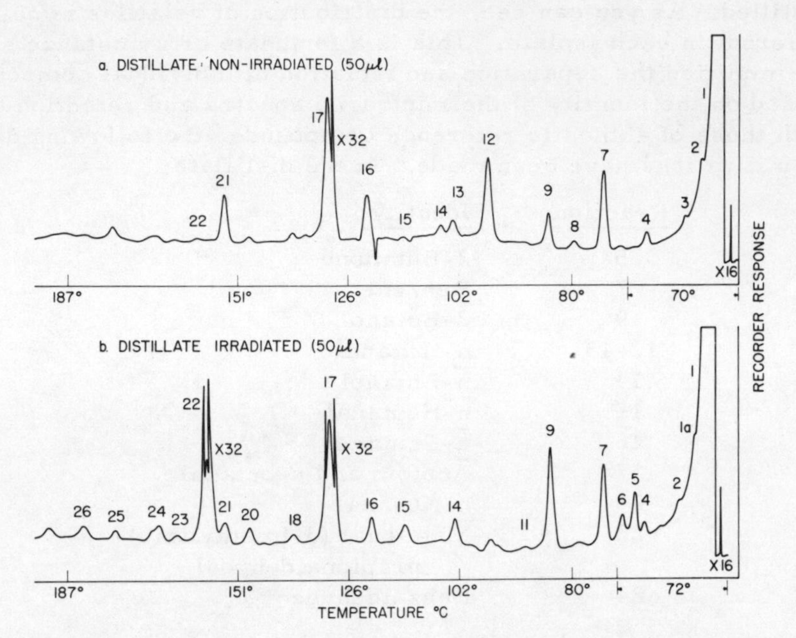

Figure 2.2. Temperature-programmed separation of isolates from nonirradiated and irradiated beef on a 20% Carbowax 20M column

and irradiated beef. It is clear that methional (fraction 22) is not present in significant quantities in nonirradiated beef. Fraction 21, an unknown alcohol, is present in both isolates but in larger quantity in the nonirradiated sample. Acetoin (fraction 17) is a major component of both isolates. There is a larger amount of n-pentanol (fraction 16) and of n-hexanal (fraction 12) in nonirradiated meat. Benzene (fraction 7) is present in both isolates and is believed to be an artifact. Ethanol (fraction 6) and 2-butanone (fraction 5) are more abundant in irradiated than nonirradiated isolates. Experience with the fractions just reviewed would lead one to expect that the spectra of fraction 9 from both irradiated and nonirradiated distillates would be identical and would be that of 2-butanol. This was not the case. Fraction 9 from nonirradiated beef contained an unknown carbonyl compound and was definitely not 2-butanol. This experience was an object lesson that one must not assume identity of fractions just because they are derived from similar isolates and have comparable retention times. Direct evidence of identity must always be obtained.

A second example of this fact is given in Figure 2.3 in which the contents of the traps from nonirradiated and irradiated beef are compared. Since fractions 5 and 9 in the irradiated isolate were shown to be hydrocarbons (n-nonane and n-decane), it would

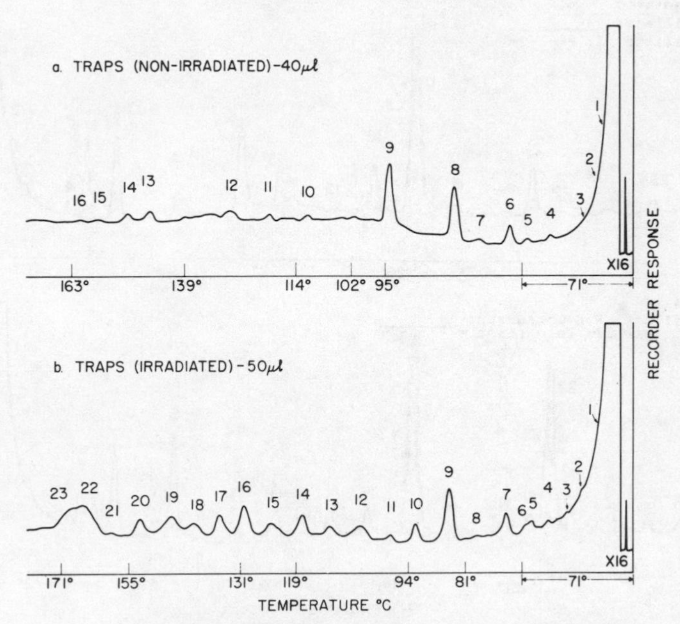

Figure 2.3. Temperature-programmed separation of isolates from nonirradiated and irradiated beef on a 20% Carbowax 20M column (comparison of contents of traps)

be logical to assume that the corresponding fractions in the nonirradiated sample contained the same substances. This was definitely not the case, as shown by rechromatography and spectra determinations. The identity of these fractions is not known. It was shown that n-hexanal was the major component (fraction 9) of the nonirradiated isolate. In the irradiated isolate the presence of n-undecane (fraction 12), 1-undecene (fraction 13), n-heptanal (fraction 14), n-dodecane (fraction 15), and 1-dodecene (fraction 16) has already been discussed.

A comparison of the composition of odor concentrates derived from concurrent and nonconcurrent procedures is made in Figure 2.4. The lower chromatogram shows a comparison of the composition of the distillate obtained by nonconcurrent radiation (the solid line) in the cobalt-60 source at 5 megarad, with that of a distillate prepared by concurrent radiation-distillation at somewhat less than 5 megarad (the dotted line). A relatively greater quantity of the saturated aldehydes was found in the concurrent preparation. This easily observed quantitative difference apparently had no effect on the sensory properties of the isolates since both exhibited strong irradiation off-odor. The n-alkanals may thus make little or no contribution to the off-odor.

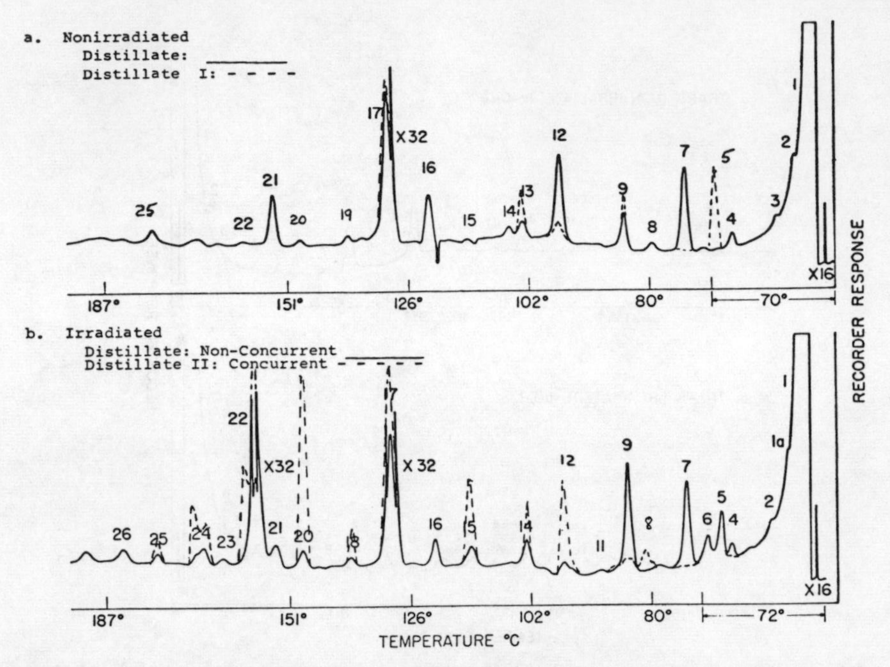

Figure 2.4. Temperature-programmed separation of isolates from nonirradiated and irradiated beef on a 20% Carbowax 20M column (concurrent and nonconcurrent procedures used)

With the evidence obtained to date from individual study of all fractions trapped, the compounds given in the last table have been identified as volatile components of enzyme-inactivated irradiated beef. Eight of the 22 compounds listed in Table 2.9 were also found in odor isolates from nonirradiated beef. On the basis of

Table 2.9. Volatile Components of Irradiated Cooked Beef

Ethanol*	Ethyl acetate
n-Butanol*	2-Butanone
2-Butanol*	n-Nonane
n-Pentanol*	1-Nonene
An unknown alcohol*	n-Decane (Tent.)
Acetoin*	n-Undecane
n-Hexanal*	1-Undecene
n-Heptanal	n-Dodecane (Tent.)
n-Octanal	1-Dodecene (Tent.)
n-Nonanal	Benzene* (Artifact?)
Methional	Benzaldehyde
Two unknown sulfur compounds	

*Also present in nonirradiated odor isolates.

current knowledge, methional and the hydrocarbons (other than benzene) are clearly products of radiation. Other components may be increased in quantity by radiation. The qualitative nature of the components has not as yet been found to be affected by the kind of irradiation or the procedure used. Radiation, whether concurrent with or followed by distillation, appears to produce essentially the same products. This may not be true, however, when the identity of the approximately 12 unknown components has been established. These unknowns were found primarily in fractions eluted at temperatures above 130°C. It is suspected that they contribute significantly to radiation off-flavor.

The contribution to radiation off-odor of the various components is not yet known. However, the fact that a rather good representation of this odor is produced when all the fractions are collected together from a temperature-programmed separation of a combination of odor isolates from irradiated distillate and traps supports our belief that the essential contributors are present.

The hydrocarbons found are believed to be bona fide products of irradiation of beef, and not artifacts from can enamel, since they were found not only in beef which had been irradiated in nonenameled cans, but also in beef which had not been irradiated in cans. The absence of mercaptans and sulfides from the components identified is explained by the evidence obtained that they were condensed in the liquid nitrogen traps which were not investigated.

Separation and identification of components of stored irradiated beef odor isolates are under current investigation. Thus, completion of the identification of the unknown volatile components of irradiated and nonirradiated beef is necessary. The next goal is to determine which components are indispensable to production of radiation off-odor and which are not. With the information thus accumulated, mechanisms for the formation of the flavor components will be postulated and tested by all means available.

References

Batzer, O. F., and D. M. Doty, 1955, J. Agr. Food Chem. 3, 64.

Batzer, O. F., A. T. Santoro, and W. A. Landmann, 1962, J. Agr. Food Chem. 10, 94; Final Project Report, Contract DA-19-129-QM-1293, File No. S-586, September 1961.

Bender, A. E., 1961, Chem. and Ind. 52, 2114.

Burks, R. E., Jr., E. B. Baker, P. Clark, J. Esslinger, and J. C. Lacey, Jr., 1959, J. Agr. Food Chem. 7, 778.

Gertner, A., and H. Ivecovic, 1954, Z. Anal. Chem. 142, 36.

Hornstein, I., P. F. Crowe, and W. L. Sulsbacher, 1960, J. Agr. Food Chem. 8, 65.

Martin, S., O. F. Batzer, W. A. Landmann, and B. S. Schweigert, 1962, J. Agr. Food Chem. 10, 91.

Merritt, C., Jr., S. R. Bresnick, M. L. Bazinet, J. T. Walsh, and F. Angelini, 1959, J. Agr. Food Chem. 7, 784.

Merritt, C., Jr., 1961, Contractor's Meeting, QM Food and Container Institute for the Armed Forces, June 1961.

Mizutani, J., and E. L. Wick, 1962, Abstracts, 142nd Meeting, American Chemical Society, September 1962, p. 15A.

Stahl, W. H., 1957, in Chemistry of Natural Food Flavors, QM Food and Container Institute for the Armed Forces, May 1957.

Wick, E. L., T. Yamanishi, L. C. Wertheimer, J. E. Hoff, B. E. Proctor, and S. A. Goldblith, 1961a, J. Agr. Food Chem. 9, 289.

Wick, E. L., J. E. Hoff, S. A. Goldblith, and B. E. Proctor, 1961b, J. Food Sci. 26, 258.

Yueh, M. H., and F. M. Strong, 1960, J. Agr. Food Chem. 8, 491.

Chapter 3

CERTAIN ASPECTS OF MICROBIAL RADIOSENSITIZING*

Gerald Silverman

Introduction

The distinction between radiosterilization and radiopasteurization is not limited to the magnitude of the radiation doses employed. In radiosterilization the objective is to completely destroy viable microorganisms, whereas in radiopasteurization a small number of organisms will survive and be capable of multiplication, given favorable conditions for growth. The behavior of these survivors is a complex subject; some aspects of it will be discussed by others in this symposium. It is of interest at this point to consider these definitions from a slightly different viewpoint – to alter somehow the radiation susceptibility of microorganisms so that dosages normally used for radiopasteurization will further drastically reduce microbial populations or even achieve sterilization. Modification of microbial radioresistivity has received considerable attention in radiobiology. The two terms commonly used to describe the changes are "protection" and "sensitization," and the tendency is to consider them as antonyms. However, this description of the mechanisms is not necessarily accurate, although the net results will be opposite – radioprotection will increase and radiosensitization will decrease microbial survival.

Radioprotection has received considerable attention, but a review of this subject is beyond the scope of this paper. The discovery that radiosensitization may be a useful tool in investigations of the biological effects of ionizing radiations is relatively recent.

The subject of radiosensitizers has been reviewed by others (R. Koch, 1957; Bridges and Horne, 1959; Eldjarn and Pihl, 1961) and should be considered to be distinct from those environmental factors which decrease radioresistivity (Stapleton, 1960). Bridges (1960, 1961, 1962) and El-Tabey Shehata (1961) reported on structurally different classes of chemicals that were capable of causing appreciable radiosensitization. A number of the observations by Bridges are of particular interest. The sensitizers

*The work herein has been performed under Grant RH-00061-08 of the National Institutes of Health.

were less active or were inactive against the test organisms when irradiation occurred in air as compared to anoxia; their effectiveness could be neutralized by cysteine, and both the chemical and test organism had to be present during irradiation for sensitization to occur.

El-Tabey Shehata (1961) found that vitamin K_5 and structurally related compounds can also cause sensitization. His work and subsequent work done with these compounds are the subject of this discussion. It should be mentioned that the term "vitamin K_5" in this context should not necessarily connote vitamin activity but is used more for convenience. I shall state at the outset that no dramatic success regarding radiosensitization of foods will be indicated, but I shall emphasize the concept of radiosensitization and the potential contributions to be obtained with this approach for understanding certain of the mechanisms of microbial radioresistivity.

The studies discussed here will be selected in a manner illustrative of the subject as a whole and with further research objectives in mind.

Time will not permit a detailed description of the experimental procedures used, and only where necessary will they be commented upon. (See El-Tabey Shehata, 1961; Silverman, El-Tabey Shehata, and Goldblith, 1962; Noaman et al., 1963; Silverman, Davis, and Goldblith, 1963.) In general, mixtures of organisms and either buffer, chemical, or food material were placed in ampoules, and the suspension flushed with air or nitrogen prior to sealing and irradiation. In certain instances a continuous gas-sparging apparatus was employed.

The terms for survival are illustrated in Table 3.1. The terms "survival fraction" (SF) and "percentage survival" (PS) are self-evident; the term "relative survival fraction" (RSF) may not be, but it is merely the ratio of the SF's obtained in chemical

Table 3.1. Terms Used for Survival

$$\text{Survival fraction (SF)} = \frac{\text{Number of organisms after irradiation}}{\text{Number of organisms before irradiation}} = \frac{N}{N_0}$$

$$\text{Percentage survival (PS)} = \frac{\text{Number of organisms in chemical}}{\text{Number of organisms in buffer}} \times 100$$

$$\text{Relative survival fraction (RSF)} = \frac{\text{SF in the presence of chemical}}{\text{SF in buffer}}$$

RSF of control samples = 1

or food to that obtained in buffer. The use of this fraction allows one to compare similar but not identical experiments performed on separate days. The RSF of control samples in the following tables will be equivalent to 1.

Table 3.2. The Effect of Vitamin K_5 on the Radiosensitivity of Microorganisms

MICROORGANISM	CONCENTRATION OF VITAMIN K_5 (PPM)	PER CENT SURVIVAL AIR	PER CENT SURVIVAL NITROGEN
ESCHERICHIA COLI	DOSE (REP)	27,000	90,000
	0	10.5	5.6
	20	10.9	0.3
MICROCOCCUS RADIODURANS	DOSE (REP)	540,000	540,000
	0	27.9	30.1
	60	5.4	2.3
PSEUDOMONAS FRAGI	DOSE (REP)	6,000	15,000
	0	6.3	32.1
	20	5.1	0.9
TORULOPSIS ROSAE	DOSE (REP)	60,000	180,000
	0	10.0	0.3
	100	1.6	0.3

El-Tabey Shehata (1961) found that vitamin K_5 was effective against such diverse microorganisms as Escherichia coli, Micrococcus radiodurans, Pseudomonas fragi, and Torulopsis rosae (Table 3.2). Two particular points are of interest in this study. For the three bacteria, the sensitizing effects were noted only in nitrogen while for the yeast, T. rosae, only in air. The apparent effect in air for M. radiodurans is due to toxicity and not to sensitization.

In subsequent work Silverman, El-Tabey Shehata, and Goldblith (1962) examined a number of compounds structurally related to vitamin K_5 for their radiosensitization against two organisms, Streptococcus faecalis (ATCC 1061) and E. coli (ATCC 9637) (Table 3.3). It was found that 4-amino-1-naphthol (4A1N) and 1-amino-2-naphthol (1A2N) were more effective radiosensitizers than K_5 and, in agreement with others (Kohn and Gunter, 1955; Howard-Flanders and Alper, 1957), that Synkavite, the diphosphoric ester of methyl naphthoquinone, did not sensitize microorganisms.

Interestingly, S. faecalis differed from E. coli in one important aspect. The increased radiolethality of K_5, 4A1N, and 1A2N was noted for S. faecalis in air and in nitrogen, although to a lesser extent in air. This led the authors to infer, by comparing the results obtained with T. rosae, S. faecalis, and E. coli, that the organism may itself influence the extent of the lethality upon exposure to radiation by unique physical or chemical interaction with the sensitizer. The effect is dependent on both dose and concentration. It was found that 4A1N is a more effective sensitizer than K_5, and that these compounds are effective at low concentrations – in the range of 10^{-4} M.

Investigations now have been completed elucidating the importance of certain substituents on the K_5 molecule that are necessary for radiosensitization. S. faecalis was the test organism

Table 3.3. Radiosensitization Effects of Compounds Structurally Related to Vitamin K_5 on Streptococcus faecalis and on Escherichia coli

COMPOUND (10^{-4}M)	RELATIVE SURVIVAL FRACTION			
	ESCHERICHIA COLI		STREPTOCOCCUS FAECALIS	
	AIR	NITROGEN	AIR	NITROGEN
	27,000 RAD	90,000 RAD	60,000 RAD	135 000 RAD
.067M PHOSPHATE BUFFER	1	1	1	1
2-METHYL-1,4-NAPHTHOHYDROQUINONE DIPHOSPHORIC ESTER TETRA SODIUM SALT (SYNKAVITE)	0.7	1.1	1.7	1.9
2-METHYL-1,4-NAPHTHOQUINONE SODIUM BISULFITE	0.4	0.9	1.6	0.06
5-AMINO-1-NAPHTHOL HCl	0.7	0.8	1.5	0.04
2-METHYL-4-AMINO-1-NAPHTHOL HCl (VITAMIN K_5)	1	0.06	0.3	0.0009
1-AMINO-2-NAPHTHOL HCl	0.3	0.007	0.002	0.00001
4-AMINO-1-NAPHTHOL HCl	0.2	0.0002	0.00002	0.00008

used. Table 3.4 presents data which indicate that the methyl group does not contribute toward radiosensitization; that 1-naphthol is only slightly active in nitrogen; and, quite unexpectedly, that 1-naphthylamine, although inactive in nitrogen, is an effective sensitizer in air. This is the only compound tested that gave results of this nature. An appreciable decrease in percentage survival is a measure of toxicity of the particular compound.

Para hydroxy- and amino-naphthalenes (Table 3.5) are very active. The activity of K_5 appears to be due to the para relationship between the hydroxy and amino groups. Activity was also noted when the amino group was in the adjacent ring, as in 5-amino-1-naphthol. Differences in toxicity appear, 1,4-diaminonaphthalene being more toxic than comparable hydroxy compounds.

Those amino- and hydroxy-naphthalene compounds which are related to 1-amino-2-naphthol (1A2N) are also active in nitrogen but not oxygen (Table 3.6). This class of compounds is the most bactericidal of the naphthalene sensitizers. Interestingly enough, beta-thionaphthol, which has a sulfhydryl group in place of the amino group of 2-amino-1-naphthol, is a protector and has a minimal toxicity.

Table 3.4. The Relative Radiosensitization Effects of Vitamin K_5, 2-Methylnaphthalene, 1-Naphthol, and 1-Naphthylamine on <u>Streptococcus faecalis</u>

OH, CH_3, NH_2 — VITAMIN K_5; CH_3 — 2-METHYLNAPHTHALENE; OH — 1-NAPHTHOL; NH_2 — 1-NAPHTHYLAMINE

	VITAMIN K_5	2-METHYLNAPHTHALENE	1-NAPHTHOL	1-NAPHTHYLAMINE
RELATIVE SURVIVAL FRACTION				
AIR	0.14	0.79	0.12	0.0000088
NITROGEN	0.00088	0.36	0.02	0.16
PERCENT SURVIVAL (UNIRRADIATED)				
AIR	89	97	87	60
NITROGEN	69	122	78	80

Table 3.5. The Relative Radiosensitization Effects of 4-Amino-1-Naphthol, 1, 4-Naphthalenediol, 1, 4-Diaminonaphthalene, and 5-Amino-1-Naphthol on <u>Streptococcus faecalis</u>

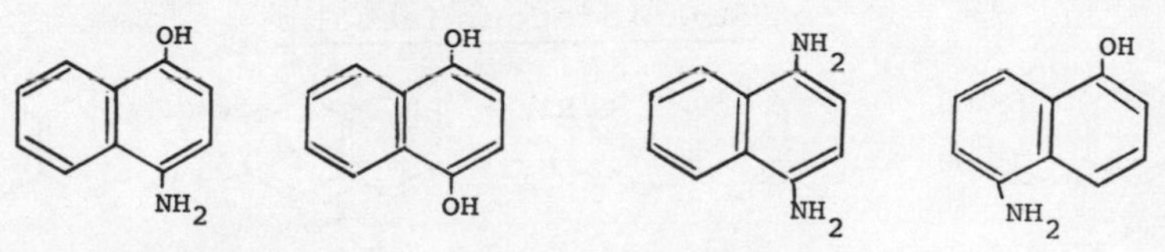

4-AMINO-1-NAPHTHOL 1,4-NAPHTHALENEDIOL 1,4-DIAMINONAPHTHALENE 5-AMINO-1-NAPHTHOL

	4-AMINO-1-NAPHTHOL	1,4-NAPHTHALENEDIOL	1,4-DIAMINONAPHTHALENE	5-AMINO-1-NAPHTHOL
RELATIVE SURVIVAL FRACTION				
AIR	0.00012	0.083	0.019	1.78
NITROGEN	0.000046	0.0001	0.00016	0.0079
PERCENT SURVIVAL (UNIRRADIATED)				
AIR	46	45	17	44
NITROGEN	47	62	32	64

Placing the amino and hydroxyl compounds in the 2 and 3 positions of naphthalene results in inactive compounds (Table 3.7) and minimal toxicity. The meta relationship, as seen in naphthoresorcinol, was also inactive.

The naphthalene structure is not essential for sensitization (Table 3.8). In fact, comparable phenyl compounds, such as hydroquinone, p-phenylenediamine, and p-aminophenol, are all active in nitrogen, although they are somewhat less effective radiosensitizers than the corresponding naphthalenes and are

Table 3.6. The Relative Radiosensitization Effects of 1-Amino-2-Naphthol, 1, 2-Diaminonaphthalene, 2-Amino-1-Naphthol, and β-Thionaphthol on *Streptococcus faecalis*

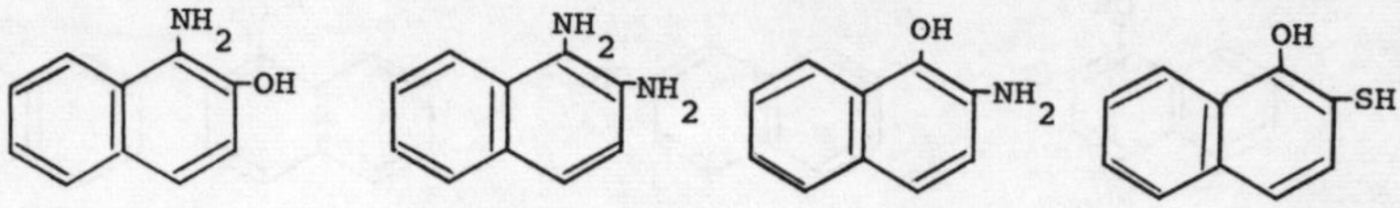

	1-AMINO-2-NAPHTHOL	1,2-DIAMINONAPHTHALENE	2-AMINO-1-NAPHTHOL	β-THIONAPHTHOL
RELATIVE SURVIVAL FRACTION				
AIR	0.0054	0.15	0.27	0.95
NITROGEN	0.000026	0.014	0.0016	1.26
PERCENT SURVIVAL (UNIRRADIATED)				
AIR	76	45	0.027	91
NITROGEN	23	2	0.14	75

Table 3.7. The Relative Radiosensitization Effects of 2, 3-Dihydroxy-Naphthalene, 2, 3-Diaminonaphthalene, 2-Amino-3-Naphthol, and Naphthoresorcinol on *Streptococcus faecalis*

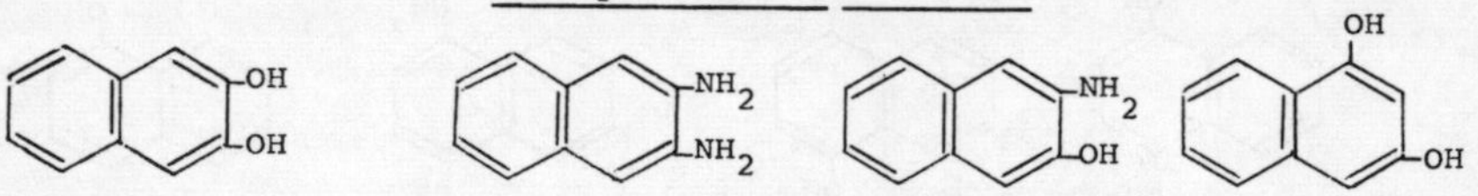

	2,3-DIHYDROXYNAPHTHALENE	2,3-DIAMINONAPHTHALENE	2-AMINO-3-NAPHTHOL	NAPHTHORESORCINOL
RELATIVE SURVIVAL FRACTION				
AIR	1.58	0.56	0.17	0.85
NITROGEN	0.0013	1.26	0.73	0.1
PERCENT SURVIVAL (UNIRRADIATED)				
AIR	93	65	73	69
NITROGEN	111	90	83	61

inactive in air and only slightly toxic.

In general, therefore, the ortho and para hydroxyl- or amino-naphthalenes are the most active radiosensitizing compounds. Since 4A1N and 1A2N are more effective sensitizers than K_5 and also more toxic, one would assume that radiosensitization is a direct reflection of bactericidal effects, but there are a number of objections to this generalization. Although most of the active compounds are more effective sensitizers in nitrogen, the largest

Table 3.8. The Relative Radiosensitization Effects of Hydroquinone, p-Phenylenediamine, and of p-Aminophenol on Streptococcus faecalis

	HYDROQUINONE	P-PHENYLENEDIAMINE	P-AMINOPHENOL
	RELATIVE SURVIVAL FRACTION		
AIR	0.5	1.21	0.15
NITROGEN	0.00043	0.00065	0.00026
	PERCENT SURVIVAL (UNIRRADIATED)		
AIR	85	84	76
NITROGEN	110	92	86

Table 3.9. The Effect of Chemical Concentration on the Radiosurvival of Streptococcus faecalis in Air and in Nitrogen

		IRRADIATED RELATIVE SURVIVAL FRACTION		UNIRRADIATED* PERCENT SURVIVAL	
COMPOUND	MOLARITY X 10^6	AIR (35,000 rad)	NITROGEN (100,000 rad)	AIR	NITROGEN
BUFFER	1.0	1.0	1.0	100	100
VITAMIN K_5	100	0.14	0.00088	89	69
	50	0.29	0.0025	104	80
	10	0.48	0.014	118	91
4-AMINO-1-NAPHTHOL	100	0.00012	0.000046	46	48
	50	0.011	0.00012	71	58
	10	0.24	0.0003	98	107
5-AMINO-1-NAPHTHOL	100	-	0.0025		72
	50	-	0.015		89
	10	-	0.036		100
	1	-	0.6		111
1-NAPHTHYL AMINE	100	0.000012	0.22	76	81
	50	0.0042	0.22	91	86
	10	0.39	0.21	95	99
IODOACETIC ACID	100	0.0000042	0.0000003	86	82
	50	0.37	0.00093	96	103

*THE INITIAL CONCENTRATION OF ORGANISMS IS 10^8 PER ML IN 0.022 M BUFFER, pH 7.

toxic effects occur in air. Many of the compounds just discussed are quite toxic but inactive in regard to sensitization.

The question of sensitization without toxicity is relative to the composition of the menstruum and the organisms employed.

Toxicity is influenced by the composition of the menstruum in which the organisms are suspended and by pH (Noaman et al., 1963). A concentration of 10^{-4} M was selected for most of this work. However, as seen in Table 3.9, sensitization occurs at lower concentrations and where little or no toxicity is evident, although this does not preclude the possibility of partial damage to the cell from which recovery is possible (Jacobs and Harris, 1960).

Bridges (1962) noted that iodoacetic acid and phenylmercuric acetate also acted as radiosensitizers. For comparative purposes the results with N-ethylmaleimide (NEM) and iodoacedic acid (IAA) are presented in Table 3.10. One other compound, p-chloromercuribenzoic acid (PCB), which has never been reported previously, also acts as a sensitizer. There is one sharp distinction between IAA, or PCB, and NEM: NEM is inactive in air.

The results with E. coli (Table 3.11) were, with two exceptions, in general agreement with those obtained with S. faecalis. Unlike most other compounds, IAA is active in air against E. coli. Furthermore, 1-naphthylamine, noted previously to be quite active against S. faecalis in air, is inactive against E. coli in either air or in anoxia. Therefore, the response to these chemicals by the test organisms varies.

It might be wise at this point to consider certain aspects of radiosensitization that need further definition. What exactly is being observed? In these examples we are observing what is of major interest — radiosurvival. In the data presented heretofore, I have employed the term "radiosensitization" in the most commonly used sense. A radiosensitizer has been defined as "a

Table 3.10. The Relative Radiosensitization Effects of N-Ethylmaleimide, p-Chloromercuribenzoic Acid and Iodoacetic Acid on Streptococcus faecalis

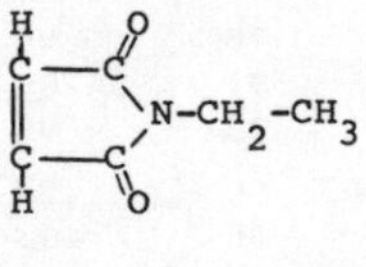

	N-ETHYLMALEIMIDE	P-CHLOROMERCURIBENZOIC ACID	IODOACETIC ACID
	(structure: HC=CH maleimide ring, N-CH_2-CH_3)	(structure: benzene ring with HgCl and C(=O)-OH)	I-CH_2-C(=O)OH
RELATIVE SURVIVAL FRACTION			
AIR	0.13	0.00074	0.0000033
NITROGEN	0.000079	0.00000025	0.00000025
PERCENT SURVIVAL (UNIRRADIATED)			
AIR	85	74	80
NITROGEN	97	61	89

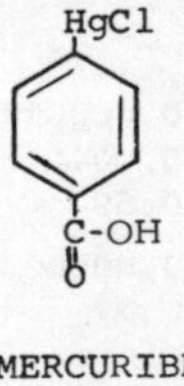

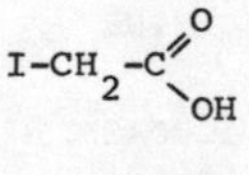

Table 3.11. The Modification of Radiosurvival of *Escherichia coli* in Air and Nitrogen by Various Chemicals

COMPOUND	IRRADIATED RELATIVE SURVIVAL FRACTION		UNIRRADIATED PERCENT SURVIVAL	
	AIR (20,000 RAD)	NITROGEN (60,000 RAD)	AIR	NITROGEN
BUFFER	1.0	1.0	100	100
4-AMINO-1-NAPHTHOL	0.16	0.00015	65	43
5-AMINO-1-NAPHTHOL	0.16	0.00015	65	62
VITAMIN K_5	0.73	0.0039	77	78
IODOACETIC ACID	0.000033	0.0027	87	130
P-AMINOPHENOL	0.73	0.013	90	94
N-ETHYLMALEIMIDE	0.95	0.05	84	100
1-NAPHTHYLAMINE	0.67	0.58	85	91

compound capable of enhancement in the radiation response accomplished by procedures acting before or during radiation exposure" (Eldjarn and Pihl, 1961). Bridges (1960) also stipulates that the sensitizer should not be toxic. As mentioned previously, although vitamin K_5 is relatively nontoxic, certain of the other compounds are toxic at the concentrations used in this study. Yet the sensitizing action does not appear to be related essentially to toxicity. In the following discussion, I should like to distinguish between sensitization and enhancement. What I should like to emphasize is that the definition of a radiosensitizer is not really satisfactory. The organism, being a biological entity, may very likely be interacting with the chemical, and the nature of this interaction is not characterized by the term "sensitization," which may not be precise enough.

For example, chemicals can alter the physiological state of the microorganism and its radioresistivity by action on the bacterial membrane (permeability, etc.), or it may be possible to disorganize or decrease the organisms' recovery mechanisms. It is known that NEM, K_5, IAA, and PCB will combine with the sulfhydryl group — in fact, NEM and PCB are reagents for sulfhydryl-containing compounds. I prefer to apply the term "radiosensitization" to those cases where the nature and extent of the ionizing track is not altered but the physiological state, or the repair mechanism, of the organisms is, and this latter alteration is reflected by a decreased radiosurvival. The term "enhancement" describes a process in which the magnitude of ionization and excitation of the lethal track is increased, but the inherent repair mechanisms of the organism are essentially unaltered. These definitions should also consider the possibility that cells exposed to radiation must not be susceptible to the compound or stress conditions which they would ordinarily survive (Silverman, El-Tabey Shehata, and Goldblith, 1962).

Table 3.12. The Minimization of the Radiosensitizing Effects of Vitamin K_5 on Streptococcus faecalis by Washing or by Dilution

COMPOUND (10^{-4} M)	IRRADIATED RELATIVE SURVIVAL FRACTION	UNIRRADIATED PERCENT SURVIVAL
BUFFER	1.0	100
CONTROL (WASHED CELLS AND CHEMICALS)		
VITAMIN K_5	0.00038	99
CELLS WASHED TWO ADDITIONAL TIMES AFTER CHEMICAL WAS ADDED		
VITAMIN K_5	0.023	85
CELLS AND CHEMICAL DILUTED 1/1000 BEFORE IRRADIATION		
VITAMIN K_5	0.11	100
CELLS WASHED OF CHEMICAL AND THEN CHEMICAL READDED BEFORE IRRADIATION		
VITAMIN K_5	0.000021	74

Table 3.13. The Effect of Combining Irradiated and Unirradiated Portions of Vitamin K_5 and Streptococcus faecalis

COMPOUND (10^{-4} M)	IRRADIATED (100,000 RAD) SURVIVAL FRACTION	UNIRRADIATED PERCENT SURVIVAL
BUFFER	0.04	100
IRRADIATION OF A MIXTURE OF CHEMICALS AND ORGANISMS		
VITAMIN K_5	0.0000032	95
ADDITION OF UNIRRADIATED CHEMICAL TO IRRADIATED ORGANISMS		
VITAMIN K_5	0.22	78
ADDITION OF IRRADIATED CHEMICAL TO UNIRRADIATED ORGANISMS		
VITAMIN K_5	0.93	135
ADDITION OF IRRADIATED CHEMICAL TO IRRADIATED ORGANISMS		
VITAMIN K_5	0.34	94

Bridges (1962) and also Dean and Alexander (1962) observed that NEM and iodoacetamide had to be present during irradiation in order to exert any effect. Moreover, their activity could be eliminated by diluting the mixture of chemical and organism prior to irradiation. Vitamin K_5 also has to be present during irradiation (Table 3.12). Essentially all of the activity is removed by washing the cells free of chemical.

Moreover, as seen in Table 3.13, the products of irradiated K_5 are not the radiosensitizing agent, nor are the irradiated cells more susceptible to the irradiated or unirradiated chemical. In agreement with others, I conclude that the organisms and the sensitizers must be present together during irradiation for sensitization to result and that this reaction is most likely to be quite rapid. Further studies are planned to determine whether or not the chemical penetrates the cell and the extent of changes, if any, in permeability.

One way to test partially whether or not vitamin K_5 and other amino- or hydroxyl-naphthalene compounds are sensitizers or enhancers is to determine their influence on the physiological state of the organisms exposed to these compounds. It can be seen in Figure 3.1 that heat resistance is reduced drastically in the presence of nitrogen. Unfortunately, there can be no correlation to radiosensitization since this effect occurs to the same extent in air and it was noted that S. faecalis is more subject to radiosensitization in nitrogen than in air. The validity of this observation will be resolved by future experiments on E. coli, where sensitization occurs only in nitrogen.

These results strongly suggest that the introduction of higher temperatures in the presence of K_5 or 4A1N may significantly increase the radiosensitization effect. In any case, until there is greater clarification of the mechanisms involved, the action of naphthalene compounds will be referred to as radiosensitizers.

I have mentioned that many of the radiosensitizers are sulfhydryl reagents. It was anticipated that sulfhydryl compounds present in foods would decrease the ability of these chemicals to sensitize organisms present in foods. El-Tabey Shehata (1961) had previously noted the decrease in effectiveness of vitamin K_5 in milk, and further work has substantiated his observation (Table 3.14). Three chemicals, K_5, 4A1N, and 1A2N, were ineffective in air and only moderately effective against the natural flora of raw milk. In an attempt to determine the cause, suspensions of S. faecalis were irradiated in pasteurized milk (Table 3.15). Moderate sensitization occurred in nitrogen, much less than in buffer at pH 7.0, but to a greater degree than the effect shown in Table 3.14, which illustrates the sensitizing effect on a mixed natural flora in raw milk. Subsequent experiments indicate that the sulfhydryl-containing whey proteins of milk are responsible for a portion of the decreased sensitization by these naphthalene

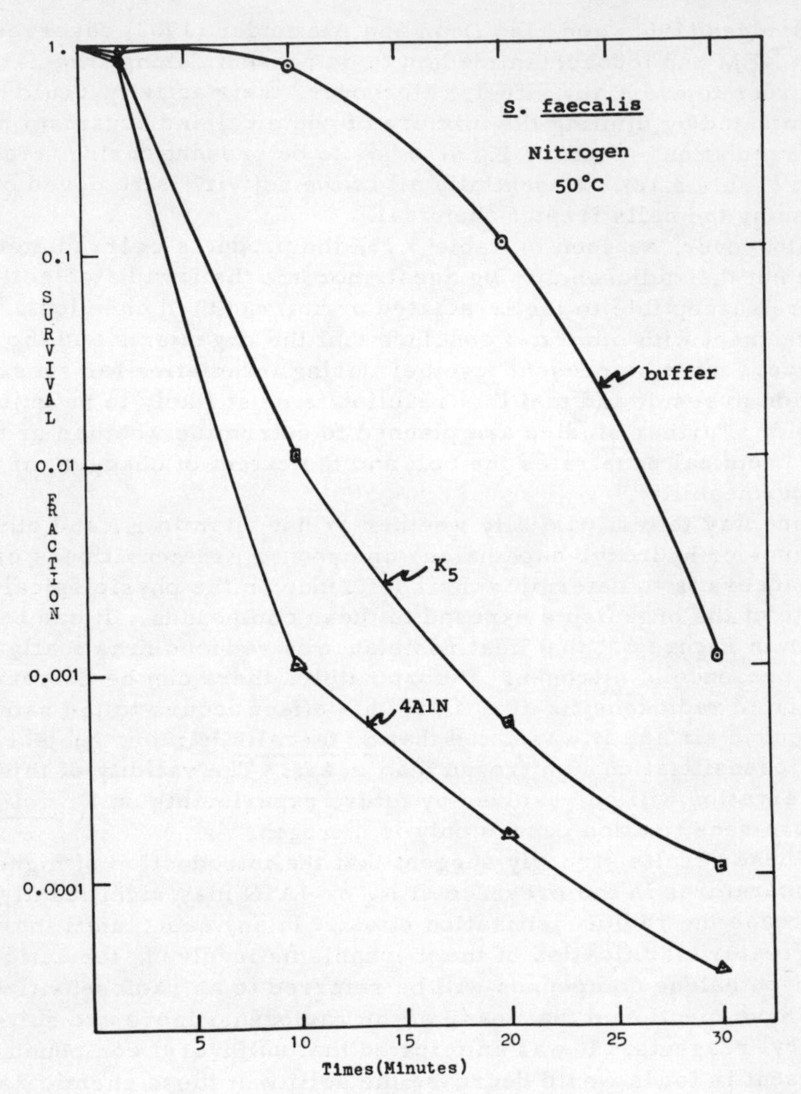

Figure 3.1. The effect of K_5, 4A1N, and buffer on the radiosurvival of S. faecalis in the presence of nitrogen

compounds (Table 3.16). Casein, a protein deficient in sulfhydryl groups, is somewhat protective toward S. faecalis. It is also seen that the activity of K_5 and 4A1N are appreciably diminished in the presence of cysteine (4A1N) more so than K_5). Two other food materials were also examined – ground beef and orange juice. No additional decrease in microbial survival over that of the controls occurred when ground beef was irradiated in the presence of K_5, 4A1N, or 1A2N. Similar results were obtained with orange juice (Table 3.17). An as-yet-unidentified yeast isolated from orange juice was inoculated into juice with and without vitamin K_5

Table 3.14. Effect of 40 ppm, Vitamin K_5, 4-Amino-1-Naphthol and 1-Amino-2-Naphthol on the Flora of Irradiated Raw Milk

	SURVIVAL FRACTION		
	AIR (20,000 RAD)	NITROGEN (60,000 RAD)	
RAW MILK	0.30	0.11	0.14
RAW MILK + K_5	0.18	0.035	0.053
RAW MILK + 4A1N	0.15	0.022	0.026
RAW MILK + 1A2N	0.18	0.076	0.065

Table 3.15. The Survival of Streptococcus faecalis in Pasteurized Milk after Irradiation in the Presence of 40 ppm of K_5, 4A1N, and 2A1N Irradiated in a Nitrogen Atmosphere

	AIR (20,000 RAD)	NITROGEN (68,000 RAD)
MILK + S. FAECALIS	0.40	0.10
MILK + S. FAECALIS + K_5	0.46	0.053
MILK + S. FAECALIS + 4A1N	0.30	0.0026
MILK + S. FAECALIS + 1A2N	0.40	0.0019

MILK – 400 ORGANISMS PER ML
S. FAECALIS – 400,000 ORGANISMS PER ML

Table 3.16. The Survival of Streptococcus faecalis Irradiated in the Presence of Casein, K_5, 4A1N, and Cysteine in a Nitrogen Atmosphere

	SURVIVAL FRACTION (80,000 RAD)
CASEIN (3%)	0.39
CASEIN + K_5 (40 PPM)	0.0026
CASEIN + 4AIN (40 PPM)	0.0014
CASEIN + CYSTEINE (28 PPM)	0.72
CASEIN + K_5 + CYSTEINE	0.068
CASEIN + 4AIN + CYSTEINE	0.32

or 4A1N. As shown, the chemicals were completely without any radiosensitizing effect. The compound 4A1N was so toxic to the yeast that irradiation was not necessary for a significant reduction of the yeast population.

In order to define further the problems inherent in the use of radiosensitizers in foods, let us look at two particularly troublesome microbial groups. One group, the microbial spore, is resistant to irradiation and to heat. As shown in Table 3.18, high

Table 3.17. The Effect of K_5 and 4A1N on the Survival of Yeast Irradiated in Buffer and Orange Juice

	SURVIVAL FRACTION	
	AIR	NITROGEN
M/60 BUFFER (pH 7.0)	0.15	0.54
BUFFER + K_5 (40 PPM)	0.00041	0.0074
BUFFER + 4AIN (40 PPM)	---	---
ORANGE JUICE (pH 3.7)	0.13	0.62
ORANGE JUICE + K_5 (40 PPM)	0.31	0.52
ORANGE JUICE + 4AIN (40 PPM)	0.25	0.79

Table 3.18. The Survival Fraction of Bacillus subtilis var niger Irradiated in K_5, 4A1N, and M/60 Phosphate Buffer, pH 7

		SURVIVAL FRACTION	
	PPM	AIR (273,000 RAD)	NITROGEN (520,000 RAD)
K_5	0	0.01	0.01
	50	0.02	0.003
	150	0.03	0.0009
4A1N	0	0.01	0.01
	50	0.03	0.003
	150	0.024	0.001
1A2N	0	0.01	0.01
	50	0.02	0.007
	150	0.03	0.003

levels of K_5, 4A1N, and 1A2N produce a moderate effect on spores of Bacillus subtilis var niger, sensitization occurring only in nitrogen and the three compounds being protective in air. Comparable results were obtained for spores of Bacillus stearothermophilus. Although the chemicals were nontoxic for B. subtilis var niger, they were toxic for B. stearothermophilus.

The other types of organism – such as M. radiodurans – are extremely resistant to radiation but not to heat or chemicals (Table 3.19). Although 150 ppm of K_5 were not toxic, comparatively low concentrations of 4A1N and 1A2N were. (Note the high dosages of irradiation required for 30 to 50 per cent survival in buffer, a level greater than for endospores.) Sensitization was evident but, except in the case of 2A1N, not sufficiently for the purposes of radiosensitization in foods. As a matter of fact, it is easier, in certain cases, to destroy these organisms with heat or by the toxic action of these compounds than by radiation. M. radiodurans presents a seemingly minor problem. This organism, which was isolated from meat, does not grow rapidly at refrigerate temperatures (Anderson, 1960). Recently another organism, a catalase-positive cocci which was isolated from fish, proved to

Table 3.19. The Influence of K_5, 4A1N, and 1A2N on the Survival of Micrococcus radiodurans Irradiated in M/60 Phosphate Buffer, pH 7

	PPM	AIR (500,000 RAD)	NITROGEN (750,000 RAD)
K_5	0	0.5	0.5
	150	0.04	0.001
4A1N	0	0.3	0.3
	10	0.06	0.02
1A2N	0	0.3	0.3
	30	-	0.0003

be as resistant as M. radiodurans when subjected to high doses of irradiation (Figure 3.2) (Davis, Silverman, and Masurovsky, 1963). The culture was found to consist of a rough and a smooth strain, the smooth strain being the most resistant.

The isolation of a second radiation-resistant organism does not mean that these organisms are widespread. To a certain extent, however, these organisms may have been overlooked in food microbiology. The new cocci grow slowly and not very well, if at all, at minimal refrigeration temperatures.

At present one can only infer that chemical radiosensitization in foods has potential applications but that success will depend upon finding chemicals and the proper conditions which will be effective for this purpose. This hope may not see fulfillment, but only continued research will elucidate this point.

I have mentioned the term "toxicity" a number of times — i.e., toxicity for microorganisms, not for humans. This is a factor which will certainly have to be resolved at the appropriate time. One other characteristic that tempers one's optimism may be the possible need, shown by many radiosensitizers, for maintaining anoxia during irradiation.

Another aspect in food irradiation that has not received the proper attention and has only been implied is the fact that the mixed microbial flora in or on a food is not distributed uniformly. Actually, there is little reason to assume even distribution. Anyone who has sampled foods for microbial counts realizes the necessity for using large enough portions for greater validity. A given food usually possesses a mixed flora, the various species growing at different rates, dependent upon the nature of the food and the temperature at which the food is stored. The result, after an interval of storage, will be discrete colonies or clones of those organisms favored, or capable of growth, under the conditions imposed during storage. Let us assume that irradiation will affect all parts with equal intensity but will require a greater number of hits to eliminate completely those areas where a higher

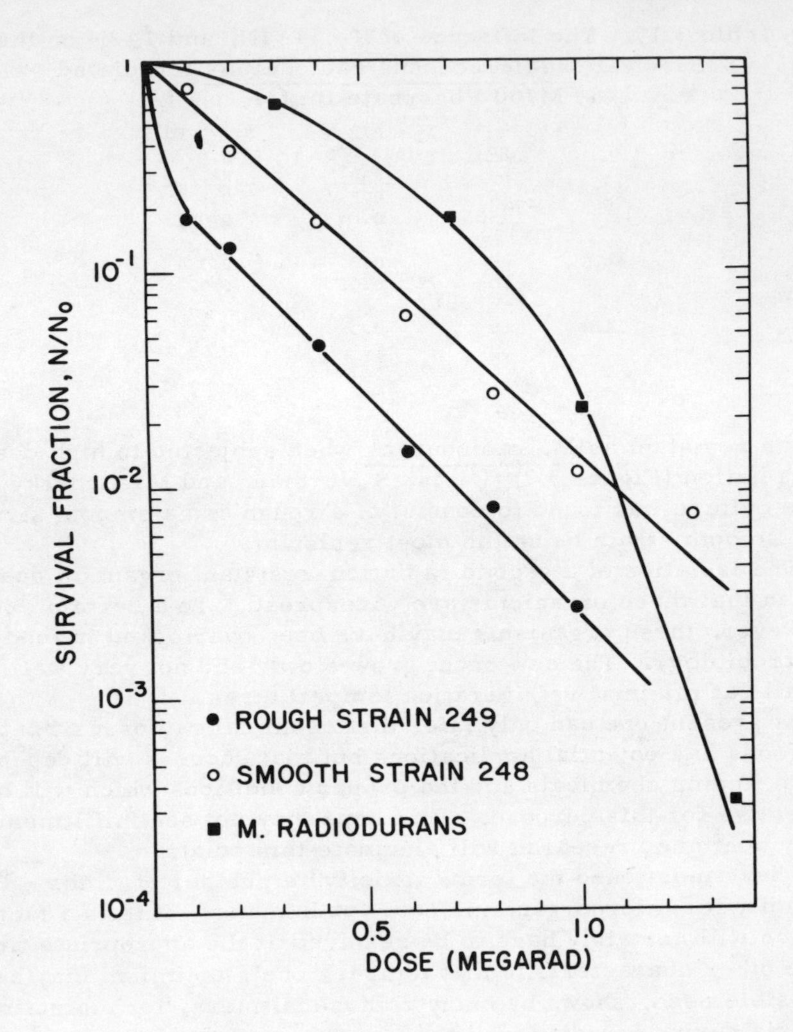

Figure 3.2. The radiosurvival of washed cells of the rough and smooth variants of the coccus and M. radiodurans

density of organisms (clones) exist, as contrasted to areas where single or few organisms are located. This analogy does not consider such important factors as the stage the organism is at, its growth cycle, the protective effect by the food constituents, or the temperature during irradiation. Regardless of these factors it becomes apparent why irradiated foods will often show a significantly greater survival of certain organisms which were capable of growth during storage than that predicted by pure culture studies where a uniform suspension is usually irradiated.

By its very nature, ionizing radiations exert their lethal action on biological targets in a random manner and can be expressed in statistical terms. This fact does not preclude the "target theory" and does not distract from the target theory's usefulness. It is likely that certain unique structures within a cell are of greater importance than others for recovery, and there are ways of increasing the effects, or efficiency, of the ionizing damage. One method is by raising the temperature during irradiation; this increases the inactivation volume (Setlow and Pollard, 1962). Radiosensitizers may act similarly, though the answer to the question of whether its radiolethality is exerted on specific cell sites or cell constituents is not known.

References

Anderson, A. W., H. C. Nordan, R. F. Cain, G. Parrish, and D. Duggan, 1956, Studies on a radiation-resistant micrococcus: I. Isolation morphology, cultural characteristics and resistance to gamma radiation, Food Technol. 10, 575.

Bridges, B. A., and T. Horne, 1959, The influence of environmental factors on the microbial effect of ionizing radiations, J. Appl. Bacteriol. 22, 96-115.

Bridges, B. A., 1960, Sensitization of Escherichia coli to gamma radiation by N-ethyl maleimide, Nature (London), 188, 415.

Bridges, B. A., 1961, The effect of N-ethyl maleimide on the radiation sensitivity of bacteria, J. Gen. Microbiol. 26, 467-472.

Bridges, B. A., 1962, The chemical sensitization of Pseudomonas species to ionizing radiation, Radiation Res. 16, 232-242.

Davis, N. S., G. J. Silverman, and E. B. Masurovsky, 1963, Radiation-resistant, pigmented coccus isolated from haddock tissue, J. Bacteriol. (in press).

Dean, C. J., and P. Alexander, 1962, Radiosensitization of bacteria with iodoacetamine, Abs. in Second Int. Cong. Radiation Res., p. 191, Harrogate, England.

Eldjarn, L., and A. Pihl, 1961, Mechanisms of protective and sensitizing action, in Mechanisms in Radiobiology (M. Errera and A. Forssberg, eds.), Vol. 2, pp. 213-296, Academic Press, New York.

Howard-Flanders, P., and T. Alper, 1957, The sensitivity of microorganisms to irradiation under controlled gas conditions, Radiation Res. 7, 518-540.

Jacobs, S. E., and N. D. Harris, 1960, The effect of environmental conditions on the viability and growth of bacteria damaged by phenols, J. Appl. Bacteriol. 23, 294-317.

Koch, R., 1957, The problem and constitution of radiation-sensitizing agents, in Advances in Radiobiology (G. C. de Hevesy, A. G. Forssberg, and J. D. Abbott, eds.), pp. 170-175, Oliver and Boyd, Edinburgh.

Kohn, H. I., and S. E. Gunter, 1955, Effect of menadiol diphosphate (Synkavite) on the sensitivity of Escherichia coli and Saccharomyces cerevisiae to X-rays, Radiation Res. 2, 351-353.

Noaman, M. A., G. J. Silverman, N. S. Davis, and S. A. Goldblith, 1963, Radiosensitization of Streptococcus faecalis and Escherichia coli, Radiation Res. (submitted for publication).

Setlow, R. B., and E. C. Pollard, 1962, Molecular Biophysics, Addison-Wesley Publishing Co., Inc., Reading, Mass.

Silverman, G. J., A. M. El-Tabey Shehata, and S. A. Goldblith, 1962, The radiosensitivity of Escherichia coli and Streptococcus faecalis as influenced by vitamin K_5 and its analogs, Radiation Res. 16, 432-440.

Silverman, G. J., N. S. Davis, and S. A. Goldblith, 1963, Modification of radiolethality by vitamin K_5 and certain analogs in model systems and in foods, Food Technol. (submitted for publication).

Stapleton, G. E., 1960, Protection and recovery in bacteria and fungi, from Radiation, Protection and Recovery, Pergamon Press, New York.

El-Tabey Shehata, A. M., 1961, Effect of combined action of ionizing radiation and chemical preservatives on microorganisms I. Vitamin K_5 as a sensitizing agent, Radiation Res. 15, 78-85.

Chapter 4

COMPLEMENTARY EFFECTS OF THERMAL AND IONIZING ENERGY*

Joseph J. Licciardello

Introduction

It has been estimated that in order to sterilize foods with ionizing radiation with the same probability of spoilage by Clostridium botulinum as is achieved by thermal processing, a dose of four to five million rads is required. Treatment of most foods with a dose of this order usually results in a severe change in color, texture, or flavor. In an attempt to diminish the radiation dose requirement, studies have been made in which the irradiation process has been coupled with some other lethal treatment such as ultrasonic energy, antibiotics, and so on. Of the various combinations investigated, the use of thermal energy with ionizing energy seems most promising. However, this discovery is not relatively new. Curran and Evans (1938) reported that exposure of certain aerobic bacterial spores to ultraviolet light sensitized the spores to subsequent heat treatment. In recent years this complementary effect between ionizing energy and thermal energy has also been found to exist for bacterial spores of food spoilage significance such as C. botulinum and C. sporogenes P. A. 3679 (Kempe, 1955; Kan et al., 1957). The complementary effect occurs only when the irradiation precedes the heat treatment.

This complementary action between ionizing and thermal energy not only is peculiar to bacterial spores, but has also been demonstrated for other forms of life and biological material such as paramecium (Giese and Heath, 1948), seeds (Smith, 1943), nucleoproteins (Giese, 1947), rats (Carlson and Jackson, 1959), and even tumor cells (Giese and Heath, 1948). It is not essential that the radiation be ionizing radiation, because the complementary effect has been observed with ultraviolet radiation (which is nonionizing) and also with visible light, providing that

*The work cited herein has been performed under Grant No. EF-6(C3) of the National Institutes of Health.

the test subjects had been treated with a photodynamic dye (Giese and Crossman, 1946).

Although the actual mechanism for the synergistic action between ionizing and thermal energy is not known, it is generally believed to be that of irradiation lowering the thermal requirement for protein denaturation. It has been proposed that irradiation may damage some of the chemical bonds which hold the main chains together in a protein molecule, and in this state upon the application of thermal energy, which causes the molecule to vibrate, it becomes disrupted (Giese and Crossman, 1946).

Since it has been well established that the extent of radiation damage to microorganisms depends on the conditions during irradiation, investigations were carried out at M.I.T. (Licciardello and Nickerson, 1962, 1963) to determine what effect the environmental condition during irradiation had on the induced heat sensitization. The test organisms studied were *C. sporogenes* P. A. 3679, an anaerobe, and *B. subtilis*, an aerobe, both of food spoilage significance.

Irradiation Prior to Heating

Figure 4.1 shows the effect of gamma irradiation on the decimal reduction time at 90°C of the spores of these organisms when the irradiation (prior to heating) was performed in air, at room temperature, and in phosphate buffer pH 7. At the time of this investigation the "rad" had not been formally accepted as the unit of absorbed radiation, and for this reason the irradiation dose is given in terms of rep.* It is readily apparent that *B. subtilis* was heat-sensitized by the irradiation to a lesser degree than was *C. sporogenes* because of the less steep slope. *B. subtilis* produces catalase, whereas *C. sporogenes* does not. Catalase can neutralize any toxic hydrogen peroxide formed within the spore during irradiation. Thus it might be conjectured that the catalase enabled the *B. subtilis* spores to survive the irradiation treatment with less radiation damage incurred than *C. sporogenes*.

The Effect of Substrate

The degree of heat sensitization induced by irradiation depends upon the substrate. In Figure 4.2 it can be seen that *C. sporogenes* was heat-sensitized to a lesser degree when suspended in a complex organic substrate such as ham purée or nutrient broth than when suspended in an inorganic medium such as phosphate buffer. Kempe (1955) also found less of a heat-sensitizing effect

*One rep is equivalent to 0.93 rad.

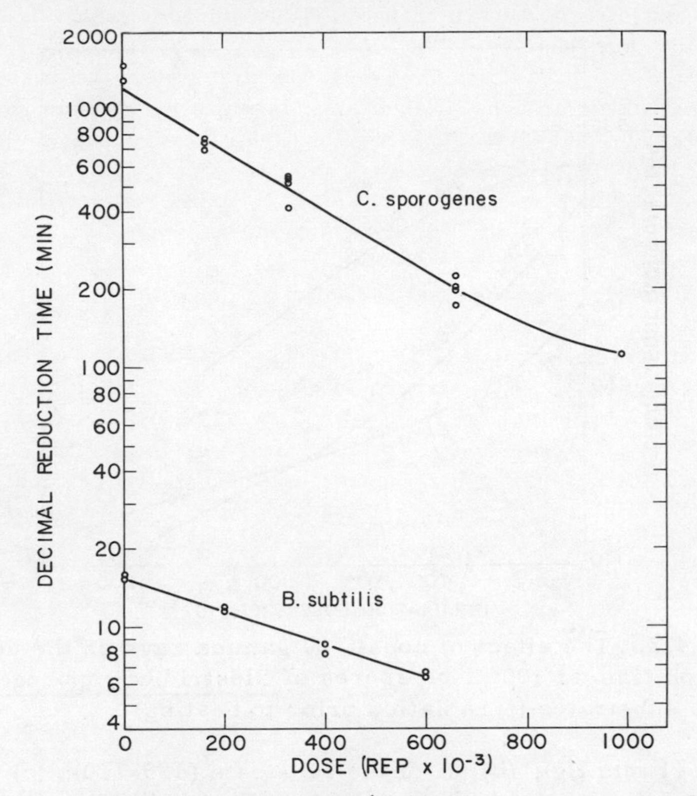

Figure 4.1. The effect of cobalt-60 gamma rays on the decimal reduction time at 90°C of spores of Clostridium sporogenes and Bacillus subtilis (irradiation prior to heating)

for spores of C. botulinum in gelatin or nutrient broth as compared with phosphate buffer pH 7. An explanation of this difference may be that in a complex organic medium there is a greater competition for the free radicals produced during irradiation and this exerts a sparing action, so to speak, on the bacteria. Or perhaps the protection by the complex substrate is afforded during the heating process.

Ham purée was selected as a substrate because, when this investigation was initiated, it was believed that if the combination of irradiation and heating proved to be a feasible method of preserving food, one potential application would be in the sterilization of whole canned hams. Recently the Danes have announced a process for producing commercially sterile canned hams utilizing both ionizing and thermal energies (Knudsen and Hansen, 1962). This process entails treating the ham with half a million rads and then heating to a center temperature of 150-158°F.

Other conditions of the environment during irradiation that were investigated included: (a) atmosphere, air versus low oxygen

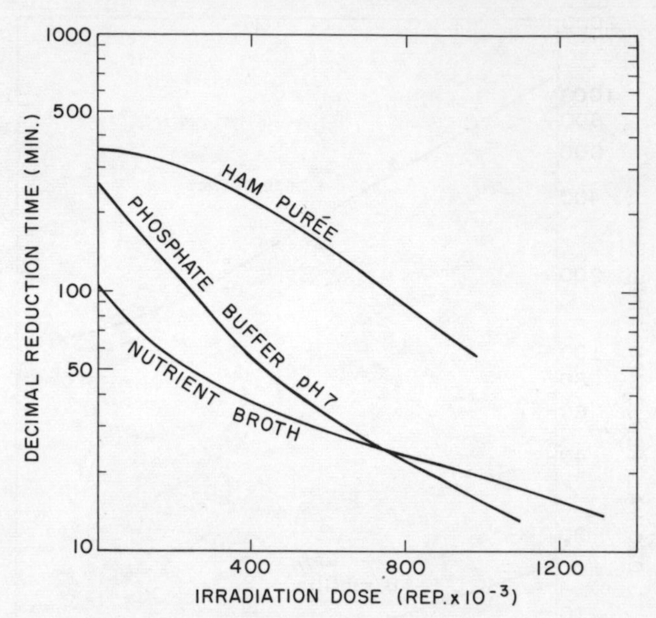

Figure 4.2. The effect of cobalt-60 gamma rays on the decimal reduction time at 100°C of spores of <u>Clostridium sporogenes</u> in various substrates (irradiation prior to heating)

tension (1 mm Hg); (b) pH of the substrate (4.5-7.0); (c) concentration of spores; and (d) temperature during irradiation. It can be stated, in general, that the degree of radiation-induced heat sensitization was not significantly changed under the different conditions of atmosphere, pH, or spore concentration employed. The degree of heat sensitization was also about the same when irradiation was carried out either in the frozen state or at room temperature (68°F), but with irradiation of <u>C</u>. <u>sporogenes</u> at about 153°F with a dose of about 600,000 rads, there was a significant increase in the degree of heat sensitization.

Simultaneous Application of Ionizing and Thermal Energies

The latter observation led to the speculation that perhaps the simultaneous application of ionizing energy and thermal energy would be more lethal than consecutive application of these two forms of energy. It would seem that if thermal energy were supplied to a bacterial cell to cause the bonds in some vital protein molecule (possibly a gene or chromosome) to oscillate at a higher frequency, then the number of radiation hits required to disrupt the molecule would be decreased while the molecule was in a state of increased vibration. This type of damage would probably be irreparable, whereas with irradiation at lower temperatures annealment of broken chromosomes or repair of damaged bonds can take place. Thus one may infer that in the process of irradiation followed by heating, under certain conditions, there can be a repair of some of the radiation damage before the thermal energy is applied.

A study has been initiated to determine the effect of irradiation at relatively high temperatures on bacterial cells and also to compare the lethal effect of simultaneous irradiation-heating with the lethal effect of irradiation followed by heating. The test organism being used for this investigation is Salmonella typhimurium suspended in liquid whole egg. The irradiation (gamma rays) is provided by a 30-kilo-curie cobalt-60 source, and the dose rate is approximately 5000 rads per minute. Some of the data that are typical of the results being obtained are presented in Table 4.1. These data show the survival of S. typhimurium after irradiation at various dose levels at 130° F, as well

Table 4.1. The Survival of Salmonella typhimurium in Liquid Whole Egg When Irradiated at 130° F, or When Irradiated at 32° F and Then Heated at 130° F

IRRADIATION DOSE (rad)	% SURVIVAL	
	IRRADIATION AT 130°F	IRRADIATION AT 32°F FOLLOWED BY HEATING AT 130°F
0	100	100
15,000	0.77	1.75
30,000	0.067	0.23
45,000	0.0047	0.059
60,000	0.00014	0.012

as survival after irradiation at various dose levels at 32° F followed by heating at 130° F for an equivalent time that was required to administer the irradiation. Thus, in both cases, the bacteria received the same quantities of thermal energy and ionizing energy except that the order of delivery was different. Note that, as the irradiation dose increased, there was a marked increase in the ratio of per cent survival of the two processes and, with a dose of 60, 000 rads, there were a hundred times more survivors when heating followed the irradiation than when heating and irradiation were carried out simultaneously.

The survival fraction was plotted as a function of irradiation dose for the two different processes, and the regression slopes were determined by the method of least squares. A comparison of the two slopes by a "t" test indicated a difference that was statistically significant at the 1 per cent level.

In Table 4.2, data of a similar nature are presented, except that in the latter case, a temperature of 100° F was employed. A

Table 4.2. The Survival of <u>Salmonella</u> <u>typhimurium</u> in Liquid Whole Egg When Irradiated at 100° F, or When Irradiated at 32° F and Then Heated at 100° F

IRRADIATION DOSE (rad)	% SURVIVAL	
	IRRADIATION AT 100°F	IRRADIATION AT 32°F FOLLOWED BY HEATING AT 100°F
0	100	100
25,000	5.50	8.40
50,000	1.20	2.42
75,000	0.29	0.72
100,000	0.055	0.244
125,000	0.009	0.053

comparison of the two different processes by the procedure outlined above showed a significant difference at the 1 per cent level.

On the basis of these preliminary results, it probably can be stated at this time that the simultaneous application of thermal energy and ionizing energy yields a significantly greater degree of destruction of vegetative bacterial cells than the process whereby the two forms of energy are administered in succession.

The survival rate during irradiation at high temperature can be partitioned into three possible components: the effect due to heat, the effect due to irradiation, and the effect due to the interaction between heat and irradiation. The component effect due to heat can be removed by determining the survival rate due to heating only at the given temperature and subtracting this heating survival rate from the over-all survival rate. By plotting the corrected survival rates obtained in this manner as a function of temperature, one can determine whether there is an interaction between heat and irradiation. This has been done for <u>S. typhimurium</u> irradiated in whole egg at 32°F, 100°F, and 130°F. The resulting graph is shown in Figure 4.3. Instead of plotting survival rate, decimal reduction dose, which is the reciprocal of survival rate and which is also the dose required to inactivate 90 per cent of the population, has been plotted as a function of irradiation temperature. Although the true shape of the actual curve cannot be established with only three points, the data nevertheless suffice to demonstrate the interaction between thermal energy and ionizing energy. More data are being accumulated to determine whether the temperature effect actually

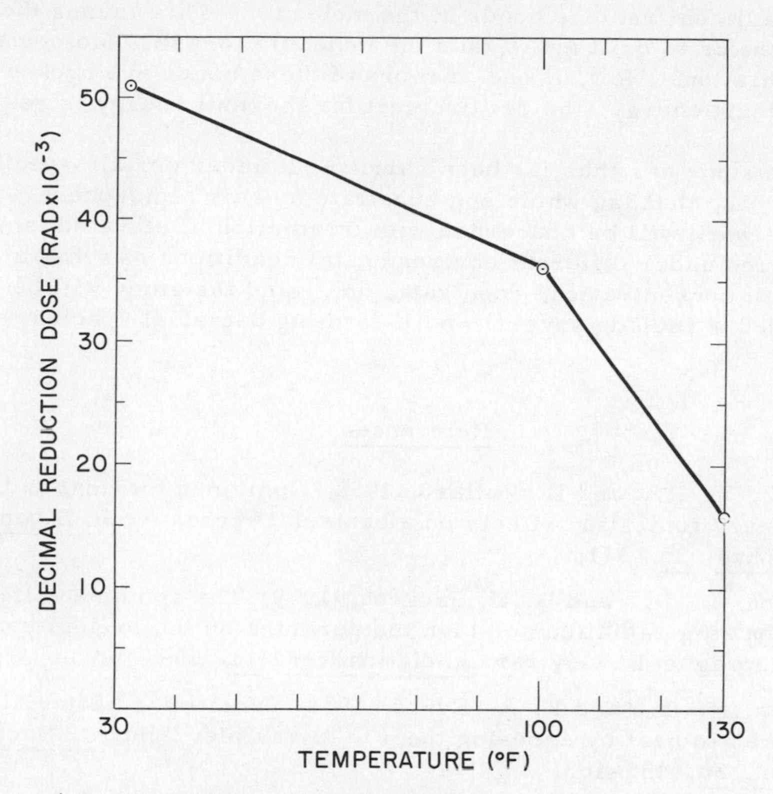

Figure 4.3. The effect of irradiation temperature on the decimal reduction dose (dose required to inactivate 90 per cent of the organisms) of Salmonella typhimurium in liquid whole egg

begins at 32°F, or at some intermediate temperature between 32°F and 100°F. Adams and Pollard (1952) irradiated a virus at various temperatures in both the wet state and the dry state. In both situations, they observed an effect of temperature during irradiation. However, this effect was apparent only from a temperature of about 110°F and above.

Various theories have been proposed to explain the increased radiosensitivity at high temperatures. These have included such ideas as: (a) increased diffusion of free radical compounds, (b) increased target size due to thermal expansion, (c) excitation that becomes important, (d) formation of new radical species at high temperature, (e) instability of molecular bonds at high temperature.

Some of these factors may play a minor role in contributing to the over-all effect. However, the model proposed by Adams and Pollard (1952) seems to be most widely accepted. They state that denaturation of protein by heat requires the rupturing of three adjacent residue bonds in the molecule. This causes the main chains to drift apart, and the molecule loses its biological configuration. Now, if one or more of these bonds are broken by ionizing energy, the requirement for thermal energy is reduced.

This study has thus far been carried out under certain specified conditions, that is, whole egg substrate and air atmosphere. Future work will be concerned with irradiation at elevated temperatures under different environmental conditions of substrate, pH, cell concentration, dose rate, etc., and the study will be extended to include several spore-forming bacterial species.

References

Adams, W. R., and E. Pollard, 1952, Combined thermal and primary ionization effects on a bacterial virus, Arch. Biochem. Biophys. 36, 311.

Carlson, L. D., and B. H. Jackson, 1959, The combined effects of ionizing radiation and high temperature on the longevity of the Sprague-Dawley rat, Radiation Res. 11, 509.

Curran, H. R., and F. R. Evans, 1938, Sensitizing bacterial spores to heat by exposing them to ultraviolet light, J. Bacteriol. 36, 455-465.

Giese, A. C., 1947, Sensitization of nucleoproteins to heat by ultraviolet radiations, Anat. Rec. 99, 672.

Giese, A. C., and E. B. Crossman, 1946, Sensitization of cells to heat by visible light in presence of photodynamic dyes, J. Gen. Physiol. 29, 193.

Giese, A. C., and H. D. Heath, 1948, Sensitization to heat by X-rays, J. Gen. Physiol. 31, 249.

Kan, B. S., S. A. Goldblith, and B. E. Proctor, 1957, Complementary effects of heat and ionizing radiation, Food Res. 22, 509.

Kempe, L. L., 1955, Combined effects of heat and radiation in food sterilization, Appl. Microbiol. 3, 346-352.

Knudsen, E. S., and P. I. Hansen, 1962, Radiation preservation of meat product past and future research at the Danish Meat Research Institute, Food Irradiation 2 (3).

Licciardello, J. J., and J. T. R. Nickerson, 1962, Effect of radiation environment on the thermal resistance of irradiated spores of *Clostridium sporogenes* P. A. 3679, J. Food Sci. 27, 211.

Licciardello, J. J., and J. T. R. Nickerson, 1963, Effect of radiation environment on the thermal resistance of irradiated spores of B. subtilis, Appl. Microbiol. 11, 216.

Smith, L., 1943, Relation of polyploidy to heat and X-ray effects in the cereals, J. Heredity 34, 131.

Chapter 5

THE STORAGE LIFE EXTENSION OF REFRIGERATED MARINE PRODUCTS BY LOW-DOSE RADIATION TREATMENT*

John T. R. Nickerson

Introduction

In 1959, the United States Atomic Energy Commission, Division of Isotopes Development, decided to investigate the possibilities of radiation preservation of food products. The Department of Nutrition and Food Science, Massachusetts Institute of Technology, was asked to make a survey to predict the probability of success of such a method of processing fishery products. This was done, and a report entitled "Evaluation of the Technical, Economic and Practical Feasibility of Radiation Preservation of Fish" was written (Proctor *et al.*, 1959).

The Atomic Energy group next requested the department to plan and outline a number of research projects pertinent to the commercial application of low-dose radiation treatment of fishery products to extend storage life at refrigerator temperatures above freezing. In a report entitled "Outline of Projects to Determine the Feasibility of Radiation Preservation of Marine Products" (Nickerson *et al.*, 1960), six different investigations were planned and outlined under the following headings:

A: Refrigerated Storage Life Extension of Radiation Substerilized Marine Products

B: The Effect of Processing (Including Radiation Substerilization) on Selected Nutritional Components of Certain Marine Products

C: Packaging Requirements for Radiation Substerilized Marine Products

D: Evaluation of Hitherto Uninvestigated Marine and Fresh Water Species for Radiation Substerilization Preservation

E: Consumer Acceptability of Radiation Substerilized Marine Products

---

*The work discussed herein has been supported by the Office of Isotopes Development of the United States Atomic Energy Commission.

F: Wholesomeness Evaluation of Radiation Substerilized Marine Products

At the present time, two of these projects have been completed in other laboratories, two are under investigation at the Massachusetts Institute of Technology and in other laboratories, and one is in the detailed planning stage.

Experimental Investigations

The project to determine the refrigerated storage life extension due to substerilization radiation of marine products is being carried out at the present time in the Department of Nutrition and Food Science at M. I. T. and is now nearing completion. This investigation has involved: (1) a survey of the bacteriological contamination of certain marine products as received from processing plants, (2) the storage life of controls and radiation-treated products held at 32 to 33° F and at 43 to 45° F as determined by counts for aerobic and anaerobic bacteria, and (3) an organoleptic evaluation of radiation-treated marine products, as compared to controls, after irradiating and holding for various periods within the storage life limit at the temperatures previously indicated.

Prior to the above investigations, initial studies on optimal bacteriological methods and media were carried out in order to ascertain those media and methods which yielded the highest counts when made on unirradiated and irradiated marine products. As a result of the above work, aerobic plate counts have been made on a medium consisting of:

BBL trypticase	15.0 g
BBL peptone	5.0 g
Bacto yeast extract	5.0 g
Sodium chloride	4.0 g
Sodium sulfite	0.2 g
1-cystine	0.7 g
Bacto dextrose	5.5 g
Bacto agar	15.0 g
Distilled water	1000 ml

Plate cultures have been incubated at 20°C for 120 hours prior to making counts.

A medium and procedures developed by Mossel et al. (1956) and modified by Angelotti and Hall (1961) have been used to make counts on anaerobic Clostridia. The medium used contained:

Bacto tryptone	15.0 g	Bacto agar	15.0 g
Bacto yeast extract	10.0 g	Distilled water	1000 ml
Iron citrate	0.5 g		

Other ingredients were sterilized separately and added to the molten medium as used so that it contained 0.05 per cent sodium sulfite, 0.12 mg/ml of sodium sulfadiazine and 10 ppm of polymyxin B sulfate.

The bactericidal agents were used in this medium to prevent the growth of aerobic-facultative types of bacteria. Clostridia grow in this medium and appear as black spherical colonies. It was shown that this medium would support the growth of *Clostridium botulinum* type E.

In the survey of bacterial contamination of marine products as prepared for fresh distribution, haddock fillets (*Melanogrammus aeglefinus*) from three producers and shucked soft-shell clams (*Mya arenaria*) from three plants were examined. Samples from each of these plants were taken at 10 to 10:30 A.M. at 12 M. and at 3 to 3:30 P.M. over a period of 28 days. Five different triplicate aerobic and anaerobic counts were made on each sample by the method previously indicated.

In order to determine the approximate storage life of radiation-treated haddock fillets and shucked clams, irradiation levels between 50,000 and 800,000 rad in increments of 50,000 rad have been used. Treated products and controls were held at 32 to 33° F and at 43 to 45° F and examined for aerobic bacteria and Clostridia over a period of time. The limit of storage (from a microbiological standpoint) was considered to be that at which either type of count indicated the presence of 1×10^6 bacteria per gram.

Organoleptic tests applied to radiation-treated (50,000 to 800,000 rad in increments of 50,000 rad) haddock fillets and shucked clams packed in hermetically sealed metal containers have been of two types. A *difference test* has been applied to determine the lowest levels of treatment at which a nonexpert panel can distinguish irradiated products from controls. A nine-point hedonic scale method was used to determine differences in irradiated products and controls, as judged by a nonexpert panel, after the irradiated product had been held following irradiation, for various lengths of time at 32 to 33° F and at 43 to 45° F. For purposes of testing, haddock fillets were cooked both by steaming and by breading with flour and deep-fat frying. Clams have been cooked only by breading with flour and deep-fat frying. A statistical analysis of all taste-test scores was made by the analysis-of-variance method (Paterson, 1939) and by Duncan's multiple-range test (Duncan, 1955) where required.

Results and Discussion

The average aerobic bacterial counts on haddock fillets from three processing plants are listed in Table 5.1. It is evident that, in any plant, average counts cannot be expected to be much lower than 400,000 per gram and usually will be much higher than this. In companies A and B there was a significant buildup in the count as the day progressed, indicating that a

Table 5.1. Aerobic Bacterial Counts – Products from Three Haddock-Filleting Plants

Company	Aerobic Counts (per gram) Average of 5 Samples for a Period of 28 Days		
	A.M.	M.	P.M.
A	410,000	450,000	880,000
B	380,000	410,000	710,000
C	1,500,000	800,000	1,100,000

Significant Differences

Company A	Morning count from afternoon count
	Noon count from afternoon count
Company B	Morning count from afternoon count
	Noon count from afternoon count
Company C	Morning count from noon count

thorough midday cleanup of equipment should have been done, but was not. In plant C, product counts were highest in the morning, indicating a totally inadequate cleanup at night. It appears that in this plant the fillets produced were actually cleaning the equipment, or wiping off the excessive bacterial contamination, until midday, after which there was again a buildup of bacteria on equipment!

Aerobic bacterial counts on shucked clams are listed in Table 5.2. There was no buildup of bacteria in the product in any of these plants as the day progressed. This is what might be expected, since the only equipment used to shuck clams is a small knife and the product is shucked directly into the final container. However, the product in plants D and E showed a count approximately 100 times greater than that of the product in plant F. Plants D and E were shucking clams dredged in Maryland and

Table 5.2. Aerobic Bacterial Counts – Products from Three Clam-Shucking Plants

Company	Aerobic Counts (per gram) Average of 5 Samples for a Period of 28 Days		
	A.M.	M.	P.M.
D	3,200,000	2,900,000	5,400,000
E	3,300,000	3,300,000	4,000,000
F	58,000	61,000	52,000

Significant Differences

No significant differences in morning, noon, and afternoon counts in any plant, but counts in plant F lower than counts in plants D and E to a highly significant degree.

brought, in the shell, to the local area by truck. Plant F shucked locally dug clams. Since there was no increase in counts during the day, indicating no plant sanitation problems, it may be concluded that bacterial counts increased in the clams as they were held and transported in the shell. It has been suggested that freshly harvested clams from Maryland may have a high bacterial count. However, it is not likely that such counts would be as high as 3×10^6 to 5×10^6 per gram. In any case this points to the need for some method of processing clams and related species which are to be handled in the shucked state to permit removal from the shell near the point of harvest and shipment in the deshelled state.

Anaerobic (Clostridia) counts are listed for haddock fillets in Table 5.3 and for shucked clams in Table 5.4. Since there were no significant differences in counts between plants or within plants for either type of product, it will merely be noted that strict anaerobes were not present in high concentrations at any time in either haddock or clams.

Table 5.3. Anaerobic (Clostridia) Counts — Haddock Fillets from Three Filleting Plants

	Anaerobic Counts (per gram) Average of 5 Samples for a Period of 28 Days		
Company	A. M.	M.	P. M.
A	130	120	160
B	43	110	79
C	120	100	120

No significant differences between plants or morning to afternoon within plants.

Table 5.4. Anaerobic (Clostridia) Counts — Shucked Clams — Three Shucking Plants

	Anaerobic Counts (per gram) Average of 5 Samples for a Period of 28 Days		
Company	A. M.	M.	P. M.
D	260	220	160
E	210	150	170
F	110	110	96

No significant differences between plants or morning to afternoon within plants.

With regard to the storage life of irradiated shucked clams and haddock fillets as determined by a maximum aerobic count of 1×10^6 per gram, extensive data will not be presented, since storage life may depend on other things than the number of bacteria present. However, storage times indicated for controls

Table 5.5. Storage Life of Controls and Irradiated Haddock Fillets and Shucked Clams as Indicated by Aerobic Bacterial Counts of Approximately 1×10^6 per Gram

Sample	Treatment	Temperature of Storage	Storage Time to Reach a Count of 1×10^6 per Gram
Haddock fillets	Control	43-45°F	5 days
Haddock fillets	Control	32-33°F	12 days
Haddock fillets	200,000 rad	43-45°F	9 days
Haddock fillets	200,000 rad	32-33°F	18 days
Haddock fillets	800,000 rad	43-45°F	30 days
Haddock fillets	800,000 rad	32-33°F	60 days +
Shucked clams	Control	43-45°F	4 days
Shucked clams	Control	32-33°F	6 days
Shucked clams	200,000 rad	43-45°F	13 days
Shucked clams	200,000 rad	32-33°F	17 days
Shucked clams	800,000 rad	43-45°F	60 days
Shucked clams	800,000 rad	32-33°F	85 days

and treated samples (low and high pasteurization doses) are listed in Table 5.5.

Organoleptic tests and bacterial counts have been made on controls and irradiated shucked clams (50,000 to 800,000 rad), the irradiated product being held for various times at 32 to 33°F or at 43 to 45°F. The results are listed in Table 5.6. It is evident that the irradiated product can be held for a considerable length of time without significant differences in organoleptic quality as judged by a nonexpert panel. Moreover, in no case after the treatments (holding temperatures and holding times indicated) was there a significant increase in the aerobic or anaerobic (Clostridia) bacterial count.

While it is obviously possible to hold irradiated shucked clams at the temperature indicated for comparatively long times after radiation treatment at sufficiently high dose levels, the storage time may be limited by factors other than bacterial growth or taste. During long storage of clams, there is an increased separation of liquid, and an obvious proteolysis occurs. Moreover, in samples packed in air a darkening of the top layer of the product occurs to an approximate depth of one-quarter inch.

The darkening of irradiated shucked clams during holding can be prevented by precooking or by the addition of sulfur dioxide. It is believed that this darkening is probably due to the enzyme tyrosinase, which is known to be present in mollusks. It is possible that sufficient sulfur dioxide to inhibit tyrosinase could be added to clams without causing detectable off-flavors.

Some of the organoleptic testing of irradiated haddock has been completed, but sufficient data are not available to report in detail.

Table 5.6. Results of Organoleptic Tests (Nonexpert Panel)—Irradiated Shucked Soft-Shelled Clams versus Controls

Irradiation Dose	Storage Temperature	Storage Time after Which Controls Not Significantly Different from Irradiated
800,000 rad	32° F	85 days
800,000 rad	43° F	40 days (test not completed)
700,000 rad	32° F	62 days
700,000 rad	43° F	50 days
600,000 rad	32° F	52 days
600,000 rad	43° F	32 days (test not completed)
500,000 rad	32° F	Not done
500,000 rad	43° F	28 days
450,000 rad	32° F	30 days
450,000 rad	43° F	17 days
400,000 rad	32° F	26 days
400,000 rad	43° F	20 days
350,000 rad	32° F	30 days
350,000 rad	43° F	17 days (test not completed)
300,000 rad	32° F	21 days
300,000 rad	43° F	19 days
250,000 rad	32° F	21 days
250,000 rad	43° F	16 days
200,000 rad	32° F	17 days
200,000 rad	43° F	13 days
150,000 rad	32° F	14 days
150,000 rad	43° F	12 days
100,000 rad	32° F	12 days
100,000 rad	43° F	Not yet done
50,000 rad	32° F	10 days (test not completed)
50,000 rad	43° F	Not yet done

Expected storage life of unirradiated shucked soft-shelled clams — maximum 6-7 days at 32° F — 4-5 days at 43° F.

It is known that if haddock is to be cooked by steaming, a significant difference in flavor between controls and product irradiated at levels higher than 350,000 rads may be detected by nonexpert taste panels. Also, regarding bacterial growth during storage at 32 to 33° F or 43 to 45° F, after radiation treatment at any particular dose level, a greater extension of storage time is obtained with shucked clams than with haddock fillets.

Conclusions

Better sanitary procedures in haddock-filleting plants would provide lower bacterial counts in the finished product and increase the storage life at refrigerator temperatures above freezing. Much lower bacterial counts in the finished product are obtained by shucking clams soon after harvesting than by transporting clams for long distances in the shell and then removing the edible portion.

Low-dose irradiation treatment of shucked clams is an acceptable process for extending the storage life at refrigerator temperatures above freezing and would facilitate the distribution of this product.

Low-dose irradiation treatment of haddock fillets will facilitate the distribution of this product in the unfrozen state. However, shorter storage times may be expected with haddock than with shucked clams after irradiation treatment.

References

Angelotti, R., and H. E. Hall, 1961, Rapid procedure for the detection and quantitative of Clostridium perfringens in foods, Proceedings of Soc. Amer. Bacteriologists, 61st Annual Meeting.

Duncan, D. B., 1955, Multiple range and multiple F tests, Biometrics 11, 1-42.

Mossel, D. A., A. S. DeBrwin, H. M. J. Van Diepen, C. M. A. Vendrig, and G. Zotelwelle, 1956, The enumeration of anaerobic bacteria, and Clostridium species, in particular, in foods, J. Appl. Bacteriol. 19, 142-154.

Nickerson, J. T. R., S. A. Goldblith, S. A. Miller, J. J. Licciardello, and M. Karel, 1960, Outline of projects to determine the feasibility of radiation preservation of marine products, Report No. 9183, U.S. Atomic Energy Commission.

Paterson, D. D., 1939, Statistical Technique in Agricultural Research, McGraw-Hill Book Co., New York.

Proctor, B. E., S. A. Goldblith, J. T. R. Nickerson, and D. F. Farkas, 1959, Evaluation of the technical feasibility of radiation preservation of fish, Report for U.S. Atomic Energy Commission, Task VI, Contract No. AT(30-1)-2329.

Chapter 6

PHYSICAL AND CHEMICAL CONSIDERATIONS IN FREEZE-DEHYDRATED FOODS*

Marcus Karel

One of the major problems in freeze dehydration of foods is the maintenance of high quality of the finished products. Freeze-dried foods have an excellent potential for good final quality but at the same time are very susceptible to several types of deterioration which occur during the various stages of processing and distribution. The various operations involved in the processing and distribution of freeze-dried foods are outlined in Table 6.1.

Table 6.1. Operations in the Processing and Distribution of Freeze-Dried Foods

Initial Quality
↓
Freezing
↓
Dehydration
↓
Packaging
↓
Storage
↓
Rehydration
↓
Final Quality

The major effects responsible for quality loss in freeze-dried foods are (Goldblith *et al*. 1963):

1. Loss of volatile flavor compounds
2. Loss of water-holding capacity
3. Oxidation of lipids

---

*The author wishes to acknowledge the cooperation of Mr. Graham Lusk and Mr. Hector Pimentel, who have obtained some of the results presented in this paper. The author is also grateful for the support of the National Institutes of Health, under whose grant, No. EF 00376-01, some of the work has been conducted.

4. Nonenzymatic browning
5. Enzymatic action

The loss of volatile flavor components is particularly important in freeze dehydration of fruits, which have a critical balance of the various organic constituents necessary for their desirable flavor. Many of the flavor and aroma constituents have vapor pressures sufficiently high to result in their volatilization and removal during the freeze drying, which is normally conducted at the rather low total pressures of 100 to 1500 microns.

The loss of these volatile components may be reduced by conducting the dehydration at very low temperatures, but this approach is neither completely effective nor economical. Reintroduction of selected flavor constituents, adsorbed on a suitable substrate, such as sorbitol, in the finished dry product may provide another approach to this problem.

Loss of water-holding capacity is a very characteristic form of deterioration of freeze-dried foods. It is usually associated with a decrease in organoleptic acceptability due to lack of "juiciness" and increase in "toughness." Typical data on this effect of freeze drying are presented in Table 6.2.

Table 6.2. The Effect of Maximum Temperature of Surface during Freeze-Drying on Rehydration Ratio and Water-Holding Capacity of Haddock Fillets

	Rehydration Ratio (g water/g solids)	Water-Holding Capacity* (g water/g solids)
Frozen control	3.9	1.25
Freeze-dried at 125°F	3.6	1.00
Freeze-dried at 175°F	3.47	0.87

*Fillets weighed before and after soaking in cool water.

**After being subjected to a uniform pressure (approximately 100 lb/in^2) between two sheets of filter paper. Moisture content of rehydrated material actually determined by analysis.

The loss of water-holding capacity is generally attributed to an increased degree of association of food polymers which are responsible for the ability of food to hold water in a tight gel structure. In plant materials these effects may be correlated with an increase in the crystallinity of the key polysaccharides: cellulose and starch. This increase in crystallinity has been shown to be associated with the very nature of dehydration, that is, the removal of water which normally maintains a separation between the individual chains, thus preventing their cross-linking.

The major mechanism of this increase in crystallinity in starch and in cellulose is thought to be hydrogen bonding.

The situation is considerably more complex in the case of proteins. In animal products the loss of water-holding capacity is thought to be due to increased cross-linking of the myofibrillar proteins, primarily actomyosin. This cross-linking may be facilitated by denaturation of the proteins, but presumably it could occur even when complete protein denaturation has not taken place. One of the methods used for the study of denaturation of actomyosin is the evaluation of the ATP-ase tion has not taken place. One of the methods used for the study of denaturation of actomyosin is the evaluation of the ATP-ase activity of this protein. Data obtained in our laboratories on the loss of ATP-ase activity of salmon, mackerel, and shrimp actomyosin are presented in Table 6.3. It is evident that loss of

Table 6.3. Retention of ATP-ase Activity in Freeze-Dried Marine Products

	Per Cent of Original Activity		
	Salmon	Mackerel	Shrimp
Dried at 125°F	72	62	80
Dried at 175°F	69	55	62
Dried at 125°F and stored 1 week at 98°F	59	50	80
Dried at 175°F and stored 1 week at 98°F	50	36	56.5

activity occurs both in drying and in storage. It should be pointed out, however, that the correlation between the enzyme activity and water-holding capacity is generally not very significant, and one is forced to the conclusion that cross-linking of myofibrillar proteins may take place in the absence of complete denaturation (Connell, 1962).

The possibility that proteins other than actomyosin may be involved in reactions leading to decreased water-holding capacity cannot be discounted. There is a need in particular to study the possible role of lipoprotein changes, since these proteins are known to be very sensitive to freezing and lyophilization. Condensation reactions between nonprotein components, especially products of lipid oxidation and reducing sugars, and proteins may also be an important source of decreased water-holding capacity (Lea, 1958).

Some of the possible mechanisms of cross-linking of food polymers are summarized in Table 6.4.

Nonenzymatic browning is another important mechanism of deterioration of dried food products. A very schematic

Table 6.4. Some of the Possible Mechanisms Resulting in Decreased Water-Holding Capacity of Foods

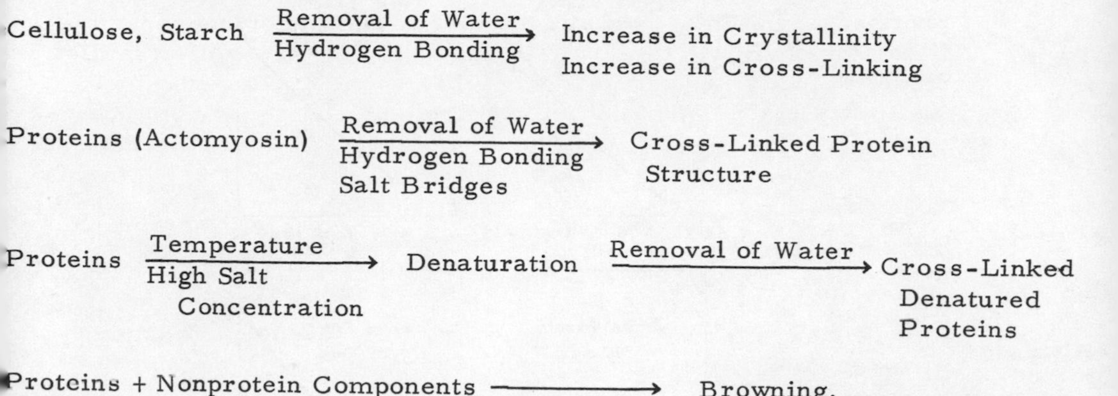

representation of the reactions involved in this type of deterioration is presented in Table 6.5. The rate of browning is very strongly dependent on moisture content; thus the contribution of browning to the deterioration of freeze-dried foods maintained at moisture contents below 2 per cent is often considered negligible. Recent work at M.I.T. has demonstrated, however, that

Table 6.5. Nonenzymatic Browning

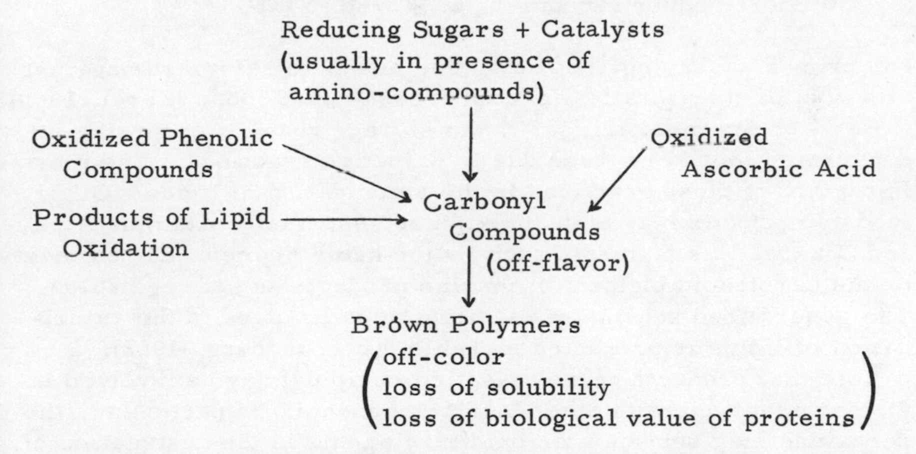

at least in some food products browning can proceed at moisture contents corresponding to less than that required for monomolecular coverage of hydrophilic sites in the food. This is illustrated in the data on dehydrated orange juice, where a finite rate of browning could be detected at moisture contents as low as 0.8 per cent moisture (Figure 6.1). Lowering of the moisture contents

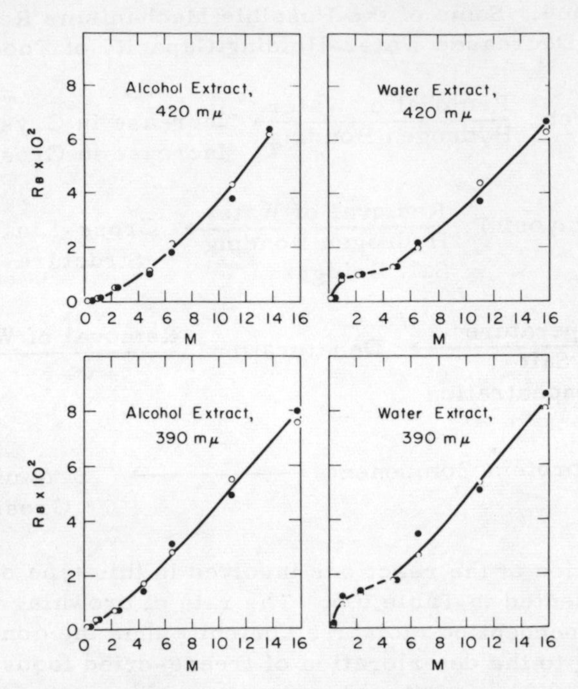

Figure 6.1. Browning of orange crystals as a function of moisture content

R_B = rate of browning (O. D. units/week)

M = moisture content (g H_2O/100 g of solids)

o Stored under vacuum • Stored in air

and proper packaging are, however, adequate safeguards against this type of deterioration in most freeze-dried foods (Karel, 1960).

Perhaps the most critical of the storage changes occurring in dehydrated foods are those due to oxidative reactions. The most important of these reactions is the autoxidation of lipids. Other food components may also be oxidized, especially vitamins A, E, and C as well as pigments such as the heme pigments of red meats or the carotenoid pigment of marine products and of vegetables. The generalized scheme of reaction steps involved in the autoxidation of lipids is presented in Table 6.6 (Lundberg, 1962).

Note that products of the oxidation of lipids may be involved in further reactions with other food components. In particular, the peroxides may serve as the oxidizing agents in the destruction of vitamins and pigments, and some of the carbonyl compounds resulting from peroxide breakdown may participate in the browning reactions and may react with various functional groups of the proteins.

Oxidation of lipids in freeze-dried foods presents a problem that is more difficult to control than similar oxidative reactions

Table 6.6. Lipid Oxidation

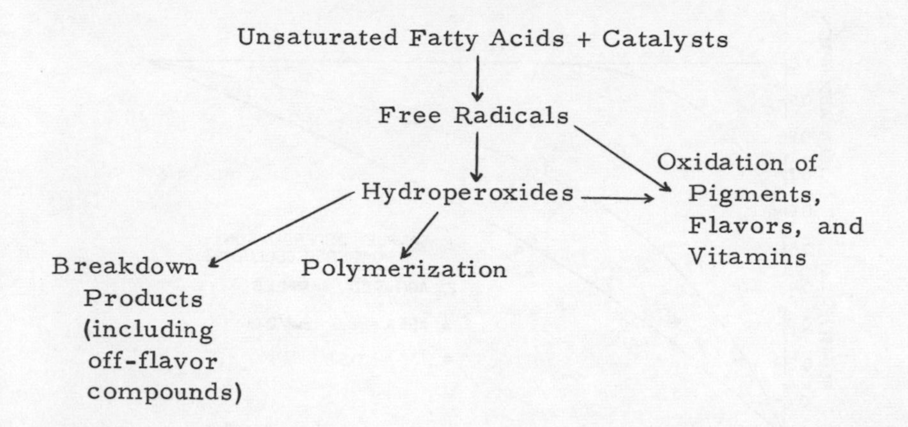

in fresh or frozen foods. There are several reasons for this difficulty.

One reason is that freeze-dried foods have a tremendous internal surface area, which allows ready diffusion of atmospheric oxygen throughout the food. Furthermore, as a result of freezing and drying, the normal distribution of lipids in the tissue may be modified, and the lipids are often distributed in monomolecular layers coating the internal surfaces. This type of lipid distribution has been shown by researchers at the U.S. Quartermaster Food and Container Institute to result in an extremely rapid oxidation. Our own experiments with model systems using linoleic acid dispersed on hydrophilic colloids not only have confirmed this accelerated rate of oxidation but have disclosed another interesting phenomenon, which is shown in Figure 6.2. This figure illustrates the dependence of oxidation rates on the partial pressure of oxygen. It can be seen that the fatty acid dispersed on a porous substrate shows almost no dependence on oxygen pressure. The practical importance of this finding is that partial reduction of oxygen concentration in the headspace of a container for freeze-dried foods will have no effect on the rate of oxidation, which will proceed rapidly until practically all of the oxygen in the container is exhausted.

The second reason for the extreme sensitivity of freeze-dried foods is the often observed phenomenon that low moisture contents promote oxidation. There are several theories for this effect of low moisture content, and active research is currently being done at M.I.T., under National Institutes of Health sponsorship, to determine which of these theories is actually applicable to the oxidation reactions in freeze-dried foods.

Finally, mention should be made of the enzymatic reactions taking place during storage and rehydration of freeze-dried foods.

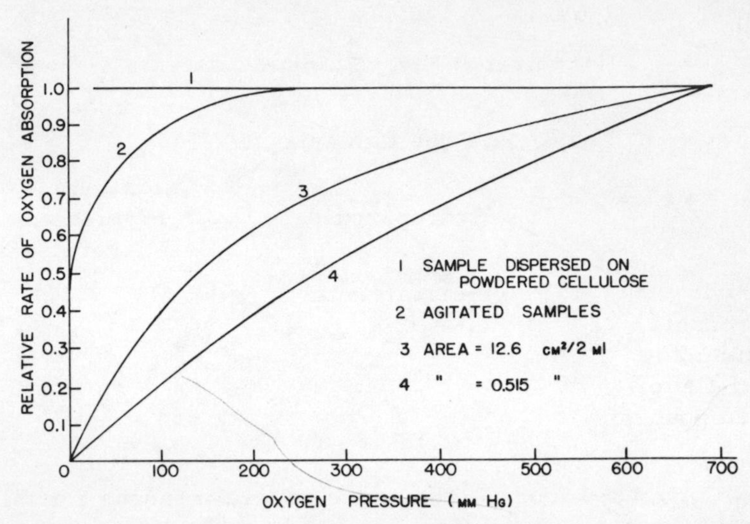

Figure 6.2. Effect of oxygen pressure on oxidation of linoleic acid oxidized under different conditions

The role of enzymatic reactions at low moisture contents has recently been reviewed by Acker (1963). The theory proposed by this author, based on studies by German investigators, is that enzymatic reactions can take place only above moisture contents corresponding to capillary condensation of water. Since most freeze-dried foods have water activities considerably below that level, one would expect that enzymic reactions in storage have no significance. There are reports in literature, however, which show that lipase in meat products, lipoxidase in some vegetable products, and polyphenolases in some fruits may be active at moisture levels as low as 3 per cent, considerably below the critical moisture contents for capillary condensation. Further research on the relationship between enzyme activity and water content is needed to clarify this matter.

Of course, even if no enzyme reactions occur during storage, extensive precautions must be taken to avoid the occurrence of such reactions during and after rehydration. Certain of the enzymes, such as polyphenolases, can cause considerable changes in color or flavor of the product even if the time interval between rehydration and consumption is relatively short.

Having examined the general nature of the physical and chemical mechanisms causing the deterioration of quality of freeze-dried foods, we may turn our attention to the effect of the individual operations in the processing and distribution of these foods. Table 6.7 shows the operational variables and possible changes during freezing. The most important changes occurring in this operation are denaturation and cross-linking of proteins, which

Table 6.7. Operational Variables and Possible Changes during Freezing

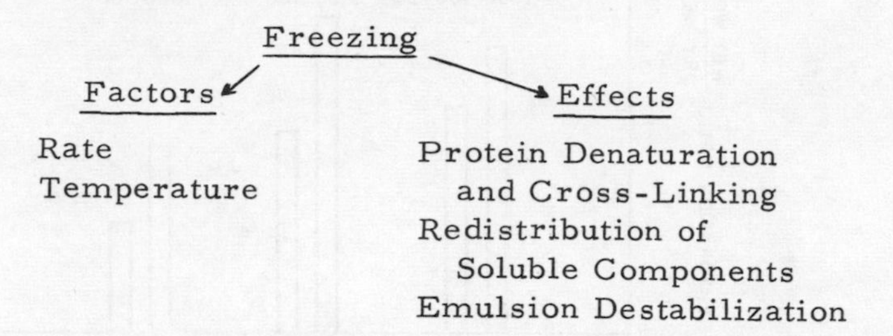

may result in impaired texture and lowered water-holding capacity. It is generally accepted that rapid freezing to very low temperatures minimizes these changes, since this method of freezing produces a very large number of small crystals, and therefore minimum mechanical damage to the tissues. In the case of freeze-dried foods, however, moderate damage to the tissues caused by intermediate rates of freezing may actually have a beneficial effect. The pores and crevices produced by such damage may facilitate mass transfer during dehydration as well as during rehydration.

These phenomena are illustrated in Figures 6.3 and 6.4, which show the effects of freezing rates on rate of dehydration, and extent of rehydration of marine products. Figure 6.3 shows the effect of freezing at 0° F and in liquid nitrogen on the rate of

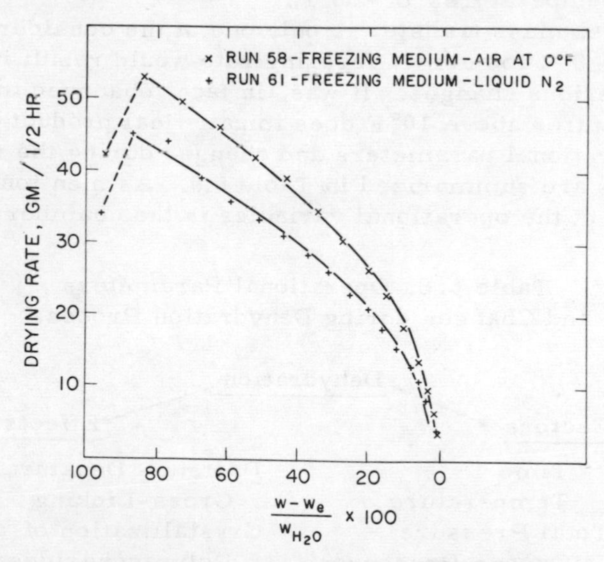

Figure 6.3. Drying rate curves for 21/25 count raw shrimp varying the temperature of the freezing medium

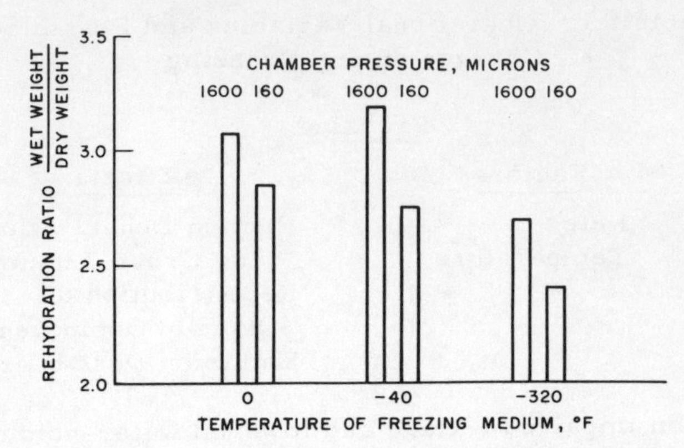

Figure 6.4. Rehydration of freeze-dried salmon

dehydration of raw shrimp. It is evident that the drying rate is faster in the samples frozen at the higher temperature.

Figure 6.4 shows the effect of the freezing medium temperature on the extent of rehydration of freeze-dried salmon, dehydrated at two different chamber pressures. The chamber pressure, it may be recalled, is the factor controlling ice temperature during the dehydration.

It is evident that freezing at moderate rates and temperatures (-40° F) results in better rehydration than freezing in liquid nitrogen. Furthermore, chamber pressures of 1600 microns, resulting in ice temperatures in excess of 0° F, were more beneficial than chamber pressures of 160 microns, which would result in equilibrium ice temperatures of -38° F.

Of course, mass transfer is only one of the considerations; it is probable that too slow a freezing rate would result in perceptible deleterious changes. It was, in fact, observed that freezing at temperatures above 10° F does impair final product quality.

The operational parameters and changes during the dehydration procedures are summarized in Table 6.8. As mentioned previously, one of the operational variables is the chamber pressure,

Table 6.8. Operational Parameters and Changes during Dehydration Process

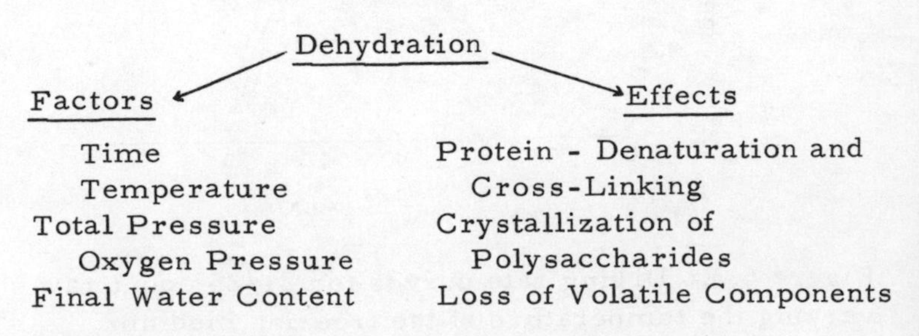
Dehydration

Factors	Effects
Time	Protein - Denaturation and Cross-Linking
Temperature	Crystallization of Polysaccharides
Total Pressure	Loss of Volatile Components
Oxygen Pressure	
Final Water Content	

which controls the equilibrium ice temperature. The other main variable under our control is the surface temperature. In most presently used procedures the heat necessary for the sublimation of ice is supplied by radiation to the surface of the dry layer, and is transferred by conduction to the surface of the frozen layer, from which the sublimation of ice takes place. Under these conditions, there exists a direct relationship between the drying rate and the surface temperature as given here:

$$r = \frac{T_s - T_i}{X_d} K_d \frac{1}{\Delta H_s} \tag{6.1}$$

where:

r = drying rate (lb $hr^{-1} ft^{-2}$)
T_s = surface temperature of dry layer (°F)
T_i = surface temperature of frozen layer (°F)
X_d = thickness of dry layer (ft)
K_d = thermal conductivity of dry layer (Btu ft^{-1} hr^{-1} $°F^{-1}$)
ΔH_s = latent heat of sublimation (Btu/lb)

Since the temperature of the ice is not an independent variable, being related to chamber pressure and dry-layer permeability, the variable most readily controlled to increase the drying rate is the surface temperature of the dry layer. In a series of experiments on shrimp, salmon, mackerel, and frozen slabs of other marine products, we have verified the validity of Equation 6.1 and found thermal conductivities of the dry layer in the range of 0.01 to 0.02 Btu $ft^{-1} hr^{-1}$ $°F^{-1}$. The effect of increasing the maximum surface temperature on the acceleration of the drying rate is shown clearly in Figures 6.5 and 6.6.

Unfortunately, increased surface temperature has a deleterious effect in accelerating some of the undesirable changes that occur during drying, especially the destruction of pigments and denaturation of proteins. For this reason, the actual temperatures

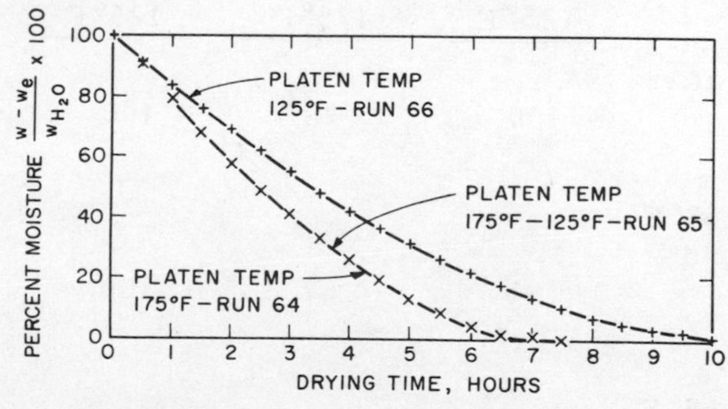

Figure 6.5. Drying curves for 1/2" mackerel steaks freeze-dried at different platen temperatures

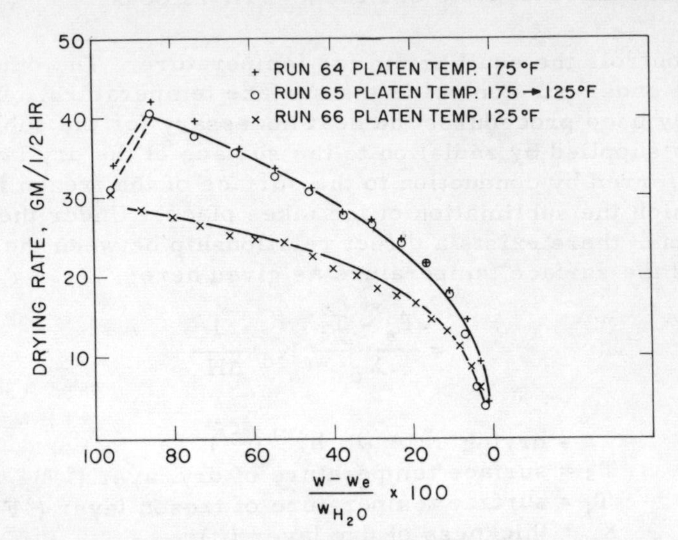

Figure 6.6. Drying rate curves for 1/2" mackerel slices varying the platen temperature

used must reflect a compromise between process efficiency and optimum product quality. High surface temperatures may have an effect not only on the quality of the product immediately after dehydration but also on the subsequent storage life. It has been observed, for instance, that the retention of the pink color of shrimp and salmon, which is due to the carotenoid pigment astacene, depends not only on conditions of storage, as will be shown later, but also on the maximum surface temperature attained during dehydration (Table 6.9). The effect is most pronounced

Table 6.9. Effect of Maximum Temperature of Surface on the Retention of Pink Color (as Determined by O. D. of Extracted Astacene) in Shrimp and Salmon

	Per Cent of Color Retention			
	Salmon		Shrimp	
	Dried at 125° F	Dried at 175° F	Dried at 125° F	Dried at 175° F
Immediately after drying	100	78	100	100
After 1 week's storage at 68° F in less than 1% oxygen	86	67	100	73
After 1 week's storage at 68° F in air	65	59	23	23

in shrimp, which have the pigment located in a thin surface layer. It may be noticed, for instance, that in this case the color is retained immediately after drying by samples processed at 125°F and at 175°F. After one week's storage in nitrogen, however, substantial loss of color occurs in samples dried at 175°F, but not in those dried at 125°F. Storage in air results in rapid color loss irrespective of the drying temperature. The data for pigment retention in salmon show a similar trend, but do not show the same tendency on a percentage basis, because the pigment is distributed throughout the tissues.

Most of the deterioration of product quality occurs during the storage period. The storage parameters and types of changes that may occur are shown in Table 6.10. It is evident that many of the reactions that take place during the initial operations of preparation, freezing, and dehydration may continue in storage. Of the storage parameters that determine the extent of deterioration, packaging protection deserves special mention. It was stated previously that both moisture content and oxygen concentration must be extremely low in order to minimize browning and lipid oxidation. Furthermore, it was emphasized that the rate of oxidation is very rapid even at low oxygen pressures, so that the only safeguard against rancidity is the limitation of the total amount of oxygen available. This brings into focus the requirement for excellent barrier characteristics in the materials chosen for packaging of the freeze-dried food. The critical nature of this requirement may be demonstrated by the following example. Assume that the product is dehydrated lean pork with

Table 6.10. Storage Parameters and Types of Changes

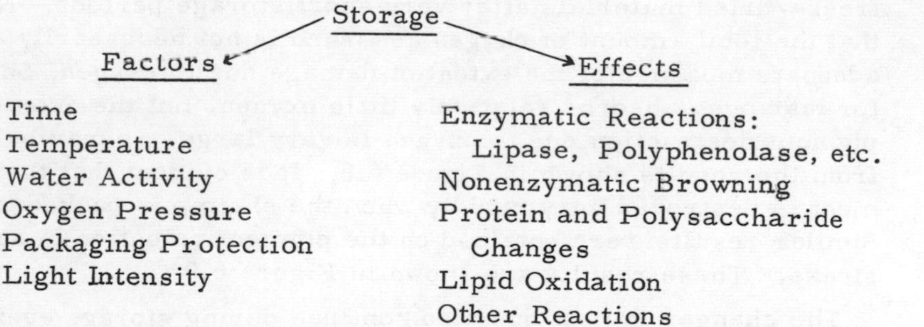

10 per cent fat on a dry basis, packed in 100-g portions in a flexible container having an area of 250 cm^2. Assuming further that the maximum acceptable peroxide value is 10 millimoles of oxygen per kg, we can calculate the allowable absorption of oxygen during the acceptable shelf life of the product as approximately 2 cm^3. The required maximum permeability of the package to keep absorption at this level is approximately 4 cm^3 of oxygen

per m^2 per atmosphere of oxygen partial pressure per day. This value is considerably below the barrier characteristics attainable with plastic films (Taylor et al., 1959). Only metal cans and completely pinhole-free laminates approach the degree of protection required. The rapidity of oxygen uptake by freeze-dried materials is demonstrated by the results presented in Figure 6.7, which

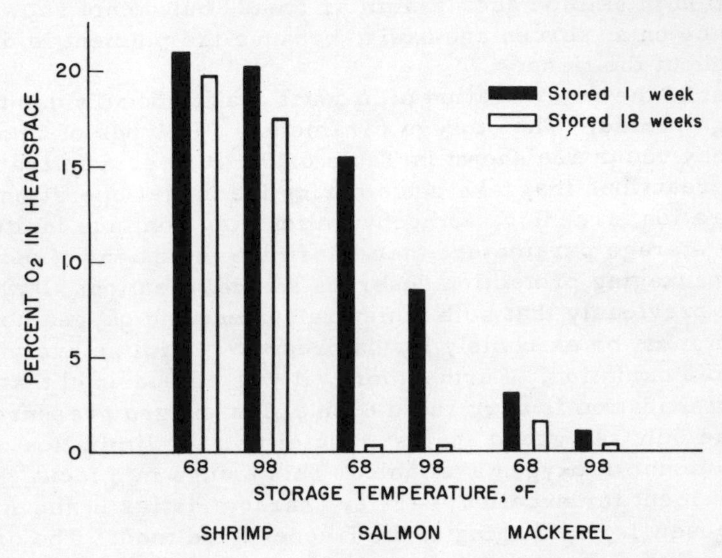

Figure 6.7. Oxygen in headspace of cans of shrimp, salmon, and mackerel packed in air

shows the oxygen remaining in the headspace of cans containing freeze-dried materials after very short storage periods. Note that the total amount of oxygen absorbed is not necessarily an adequate measure of the extent of damage due to oxygen. Shrimp, for instance, absorbs relatively little oxygen, but the extent of pigment destruction due to oxygen is very large, as may be seen from the results shown in Figure 6.8. It is evident that the pigment is destroyed very rapidly when the shrimp is packed in air. Similar results were obtained on the pigment retention in salmon steaks. These results are shown in Figure 6.9.

The changes in proteins also continue during storage even at extremely low moisture contents. Figure 6.10 shows the decrease in ATP-ase activity of shrimp, salmon, and mackerel during short-time storage under nitrogen. The products were dehydrated at 125°F to a final moisture content of less than 2 per cent. It is evident that, even when conditions of dehydration and storage are mild, there is a significant continuation of protein denaturation.

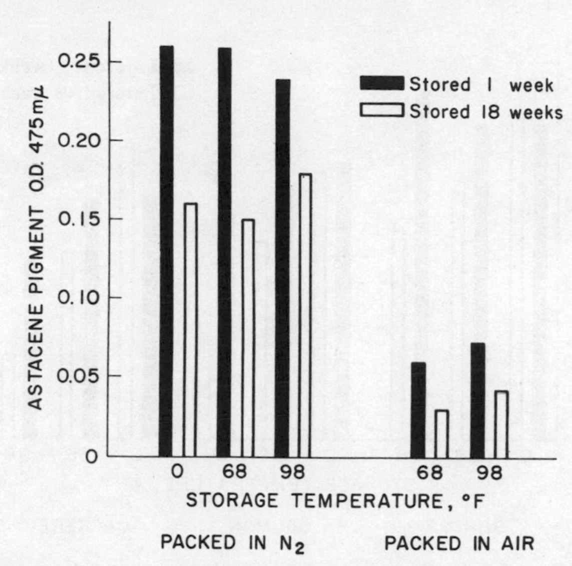

Figure 6.8. Astacene pigment in shrimp, freeze-dried at 125° F

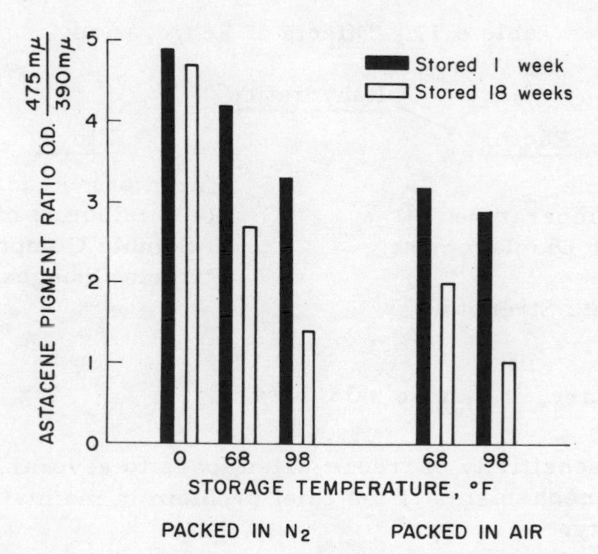

Figure 6.9. Astacene pigment in salmon freeze-dried at 125° F

The final step in the life cycle of freeze-dehydrated foods is their rehydration. Table 6.11 summarizes the effects of this operation. One of the factors that we investigated in the laboratory was the influence of the temperature of the rehydration water. We found that rehydration at high temperatures results in impaired quality, owing probably to its effect on proteins.

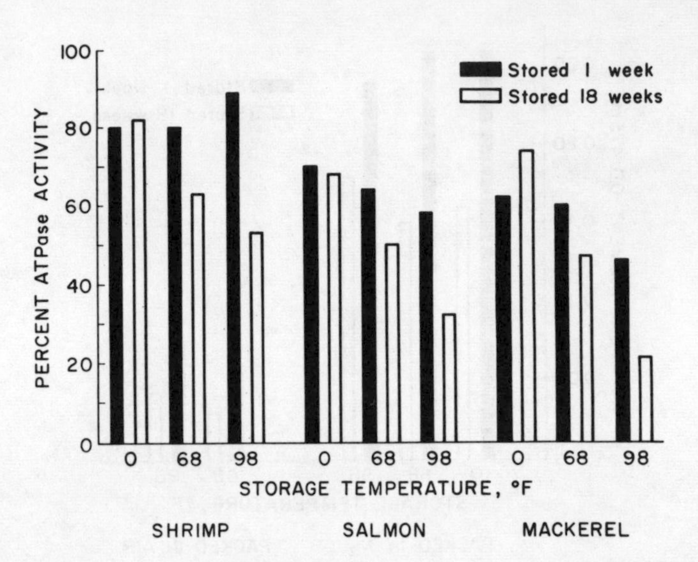

Figure 6.10. ATP-ase activity of shrimp, salmon, and mackerel freeze-dried at 125° F

Table 6.11. Effects of Rehydration

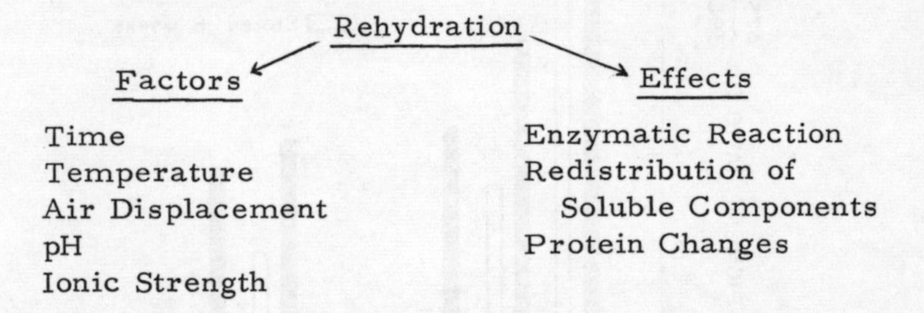

Rehydration

Factors	Effects
Time	Enzymatic Reaction
Temperature	Redistribution of Soluble Components
Air Displacement	Protein Changes
pH	
Ionic Strength	

In summary, it may be said that:

1. The sensitivity of freeze-dried foods to several deteriorative mechanisms is the chief problem in maintaining good quality.

2. The nature of the deteriorative reactions has been greatly clarified by research in the past decade, but many of the reaction paths and factors affecting these reactions require further research. This is particularly true in the case of protein changes and oxidative reactions at low moisture contents, as well as in the case of certain off-flavor problems the nature of the development of which is entirely unknown.

3. Until specific means of preventing the occurrence of deteriorative reaction are developed, the only path to good quality is the control of variables in the various operations with a view to minimizing the extent of deterioration. In this connection, it must be recognized that variables in a given operation may have an effect not only on immediate deterioration, but also on deterioration in subsequent operations.

References

Acker, L., 1963, Enzymic reactions in foods of low moisture content, Advan. Food Res. 11.

Connell, J. J., 1962, The effects of freeze drying and subsequent storage on the proteins of fresh foods, in Freeze-Drying of Foods, NAS-ARC Symposium, Washington, D.C.

Goldblith, S. A., M. Karel, and G. Lusk, 1963, The role of food science and technology in freeze dehydration of foods, Food Technol. 17 (2); (3) 22.

Karel, M., 1960, Some effects of water and of oxygen on rates of reactions of food components, Ph.D. Thesis, M.I.T.

Lea, C. H., 1958, Chemical changes in the preparation and storage of dehydrated foods, in Fundamental Aspects of Dehydration of Foodstuffs, Soc. Chem. Ind., London.

Lundberg, W. O., 1961, Autoxidation and Antioxidants, Vol. I, John Wiley and Sons, New York.

Taylor, A. A., M. Karel, and B. E. Proctor, 1960, Measurement of oxygen permeability, Mod. Packaging 33 (10), 131.

Chapter 7

MICROBIOLOGICAL CONSIDERATIONS IN FREEZE-DEHYDRATED FOODS*

Samuel A. Goldblith

Introduction**

The title of this paper is certainly an all-inclusive one and presupposes a great deal of knowledge, possession of much data, and wisdom on my part. As the day of this symposium drew near, I began to realize the enormity of the task, the paucity of the data, and the lack of wisdom. I recognize full well the myriad of information on the microbiology of freeze-dried foods now available in the freeze-dehydration industry — probably far more extensive data than those to be presented here.

However, I would like to consider some of the microbiological factors which I feel may be important in the freeze dehydration of foods and to present some preliminary data indicative of the program in this field under way in our laboratories.

In the microbiological evaluation of any "new" food process or foodstuffs processed by a "new" method, consideration might well be given to the following:

1. Are optimal methods of microbiological analysis now available, or must new ones be developed which will reflect the true microbial "picture"?

2. Does the process create "metabolic" injury of the contaminating organisms causing possible mutations, and are these mutations of public health significance?

3. Does the freeze-dehydration process itself destroy bacteria (and to what extent) as do other methods of food processing?

---

*The work cited herein has been carried out under Grant No. EF-00314-02 of the Public Health Service.

**The work discussed herein is from data obtained by T. Sinskey, I. Pablo, and A. McIntosh (graduate students in this department working with Dr. Gerald Silverman and the writer), to whom the writer expresses deep gratitude for permission to quote some of these preliminary findings.

4. Can or should microbiological standards be developed which will reflect a true picture of the potential microbiological hazards, if there be any, of the raw materials and of the manufacturing practices employed?

In any consideration of the microbiology of freeze dehydration, one should particularly recall that freeze dehydration is the technique of choice for the preservation of cultures of bacteria, yeasts, molds, and viruses. Thus, the microbiological considerations of this particular method of food processing are generally of particular significance.

I shall attempt herein to confine my remarks to the above points and to present data illustrating some of these points. Since, as I have said earlier, these are but preliminary data, they should be regarded as but a progress report on the role being played by our laboratories in this new and exciting method of food preservation.

Determination of Optimal Medium for Culturing Bacteria in Freeze-Dried Foodstuffs

Obviously, no single microbial species will respond quantitatively to the same degree to a given culture medium as another species of organism. Undoubtedly, however, many related species do respond, grow, and multiply to the same degree to a given, carefully selected culture medium; and upon this fact rests our present system of quantitative estimation of the microbiological content of foods.

As one of the early experiments in the program in our laboratories relating to the microbiological considerations in freeze-dehydrated foods, the resolution of the problem of optimal medium and diluent for a simple standard plate count was felt worthy of immediate attention.

Inasmuch as Dr. Silverman, Dr. Nickerson, and their co-workers had already conducted survey studies on frozen sea food products (Silverman, 1961a, b) and had done extensive work on optimal media, etc., comparative studies on frozen raw shrimp and freeze-dried shrimp were initiated first.

The first problem studied was that of optimal culture medium. Table 7.1 presents comparative data on the standard plate count of frozen raw shrimp using five different agars. In addition, comparisons were made on the use of trypticase diluent *vs.* distilled water in the disintegrating blendor and on the use of trypticase diluent *vs.* distilled water for the dilution blanks (Table 7.1).

These data are fascinating as well as significant. In all instances, regardless of which plating agar was employed or whether disintegration took place in the presence of trypticase diluent or distilled water in the blendor, much higher counts were obtained when trypticase diluent was used for dilution blanks (column 4, Table 7.1). This experiment was repeated, with similar results.

Table 7.1. Effect of Diluent and Plating Medium on the Standard Plate Count of Frozen Raw Shrimp

	AVERAGE STANDARD PLATE COUNT (NO. OF BACTERIA/G $\times$ 10^{-4})			
	TRYPTICASE DILUENT IN BLENDOR		DISTILLED WATER IN BLENDOR	
MEDIUM	DIST. WATER	TRYP. DIL.	DIST. WATER	TRYP. DIL.
EUGON AGAR	36	145	54	197
EUGON + 0.5% YEAST EXTRACT AGAR	25	131	22	163
PLATE COUNT AGAR	31	131	26	178
TRYPTICASE SOY AGAR	36	157	25	203
TRYPTICASE SOY + 0.5% YEAST EXTRACT AGAR	31	161	28	214

INCUBATION TEMPERATURE 20°C FOR 5 DAYS.
EACH FIGURE REPRESENTS AVERAGE OF 5 PLATES.

While differences in counts were not very great among the different agars, the highest counts were obtained when trypticase soy agar containing 0.5 per cent yeast extract was used.

Table 7.2 presents, in summary form, data indicating that use of trypticase soy agar (+0.5 per cent yeast extract) for freeze-dried shrimp will result in higher counts (although not tremendously so) than the other plating media which were studied.

Table 7.2. Effect of Plating Medium on the Standard Plate Count of Freeze-Dried Raw Shrimp

MEDIUM	AVERAGE STANDARD PLATE COUNT (NO. OF BACTERIA/ G $\times$ 10^{-3}) TRYPTICASE DILUENT IN BLENDOR AND TRYPTICASE DILUENT
EUGON AGAR	690
PLATE COUNT AGAR	650
TRYPTONE GLUCOSE EXTRACT AGAR	650
TRYPTICASE SOY + 0.5% YEAST EXTRACT AGAR	760
VEAL INFUSION AGAR	730

INCUBATION TEMPERATURE, 20° C FOR 5 DAYS
EACH FIGURE REPRESENTS AVERAGE OF 3 PLATES IN EACH OF 3 BLENDORS (TOTAL OF 9 PLATES)

Obviously, these data on freeze-dried shrimp require confirmation, and studies are needed on the determination of optimal media for other freeze-dried foodstuffs. These data do indicate, however, insofar as standard plate counts are concerned, the similarity of response of freeze-dried raw shrimp and frozen raw shrimp.

Insofar as cooked shrimp are concerned, similar comparative studies are also needed, since, by and large, cooked shrimp are essentially sterile immediately after cooking, and the microbial flora which develop on these products usually are of human origin and differ from the normal flora of raw shrimp.

Temperature of Incubation

As might be expected, incubation of both freeze-dried raw shrimp and frozen raw shrimp resulted in higher counts being obtained at 20°C than at 30°C or 37°C (Table 7.3).

Table 7.3. Effect of Temperature of Incubation

	Frozen Raw Shrimp (No. of Bacteria/g)	Freeze-Dried Raw Shrimp (No. of Bacteria/g)
20°C	1.67×10^6	7.6×10^5
30°C	1.32×10^6	3.9×10^5
37°C	0.22×10^6	1.2×10^5

Trypticase soy agar + 0.5% yeast extract and trypticase diluent

Incubation time: 20°C - 5 days
30°C - 3 days
37°C - 2 days

Again, while these data are only preliminary, relatively few in number, and on only one foodstuff, they indicate normal microbial response to the particular temperature for the particular foodstuff.

Model Systems

Much of the information on the chemical and physicochemical changes and kinetics thereof has been obtained from studies on model systems (Karel, 1960). It was felt that valuable information might be obtained on the use of relatively simple model systems (simple in comparison with complex foodstuffs) inoculated with a single species of bacteria.

To this end, gels made from 2 per cent gelatin with and without the addition of 6 per cent dextrose were inoculated with measured amounts of washed cell suspensions of *Streptococcus faecalis* and *Pseudomonas perolens*. These gels were frozen and then freeze-dried in a Virtis drier for 15 hours. The final moisture content of the dried gelatin gels was 1.1 per cent, whereas that of the gels containing added dextrose was 0.22 per cent.

Aliquots of the dried gels were then put into storage under nitrogen and under air. These storage experiments have just been initiated; and though no storage data on loss of viability in storage under oxygen *vs.* nitrogen atmospheres are available

Table 7.4. Model Systems, 2% Gelatin *vs.* 2% Gelatin + 6% Dextrose

Species	2% Gelatin (Survival %)	2% Gelatin + 6% Dextrose (Survival %)
S. faecalis	30.2	26.8
P. perolens	0.2	69.0

yet, Table 7.4 presents data on the effect of freeze dehydration on the survival of these two species following freezing.

While the lyophilization process affects to some extent the numbers of organisms surviving, *Pseudomonas perolens* appears to be markedly affected in the case of the gelatin base alone. However, with 6 per cent dextrose, fairly good survival is obtained. Whether this is an artifact or not is as yet not known. These experiments are being repeated. Suffice it to say, however, that the addition of dextrose makes little, if any, difference in the case of *Streptococcus faecalis* but certainly appears to enhance the survival in the case of *Pseudomonas perolens*. The storage data should prove interesting; and by studies on such relatively simple systems it is hoped that quantitative data of these types may offer some clues as to the mechanism of survival and destruction during lyophilization.

Metabolic Damage

It is not only of theoretical interest but also of practical significance to ascertain whether the lyophilization process causes mutagenic effects to take place. It is readily apparent that if such changes do take place, some species of nonpathogenic organisms might conceivably become pathogenic, and vice versa. This implication, of course, is of tremendous public health significance.

Preliminary studies have been initiated to determine whether freeze dehydration causes "metabolic damage," using *Salmonella typhimurium* inoculated onto seared round beef cut out with sterile cork borers. The "wet" meat was inoculated with approximately 53 to 86 $\times 10^6$ organisms per gram (uninoculated controls were found to have approximately 1800 to 2700 organisms per gram, mostly psychrophiles).

These samples were then cultured on both a "complete" medium and a "minimal" medium in order to determine metabolic damage (Table 7.5).

Straka and Stokes (1957) have defined "metabolic" injury as the ability of the particular organism to grow on a complete medium and not on a minimal medium. The complete medium used herein was trypticase soy agar containing 1 per cent yeast extract, and the minimal medium was that of Davis (1950).

Table 7.5. "Metabolic" Damage, Survival of S. typhimurium Inoculated onto Lean Beef

	Bacterial Counts ($\times 10^4$)	
	Trypticase Soy Medium + 1% Yeast Extract	Minimal Medium
Average Count	150	89

$$\% \text{ undamaged} = \frac{89}{150} \times 100 = 59\%$$

$$\% \text{ damaged} = 41\%$$

Table 7.5 presents, in summary form, the data obtained and indicates "metabolic" injury to the extent of approximately 41 per cent. (Some four runs were made with good replication.)

Biochemical reactions of some of the isolates have been obtained, and very preliminary data (and I emphasize the paucity of the data available at this time) indicate some mutagenic effects. These data are but few in number and need extension both as to the quantitative degree of mutants produced and as to whether the toxic properties of the mutants have been affected. I might add that, of the 54 colony isolates from the freeze-dried samples upon which biochemical reactions were studied, all still showed positive serological agglutination tests to polyvalent antisera, although 8 atypical sugar fermentation mutants were obtained. Certainly these data are preliminary but indeed are worthy of extension in view of the tremendous public health implications. In addition to the possible "metabolic" injury induced by the freeze-dehydration process itself, attention will be given to the effects of storage conditions — air vs. nitrogen, moisture content, temperature of storage, etc. — upon the types and numbers of organisms surviving the process.

Standards for Freeze-Dried Foods

Inevitably, in a symposium of this type, the question of microbial standards for freeze-dried foods will crop up.

Obviously, for meaningful standards to be set up, the following a priori factors should be carefully considered in order to arrive at sound judgments:

1. The standards should reflect the degree of potential public health hazard.

2. The standards are meaningful and enforceable.

3. The standards reflect the degree of "good manufacturing practice" employed.

These factors, of course, presuppose knowledge of proper methodology for the particular freeze-dried foodstuff. This entire subject has been well covered in a masterful manner by Levine (1961). The preliminary data presented herein indicate that, to this end, merely a start has been obtained on one product — freeze-dried shrimp.

The next and obvious task is to do similar studies on methodology with other freeze-dehydrated foodstuffs as test materials and to compare the results obtained on these materials with data on fresh and frozen counterparts.

This discussion of "standards" brings up some of the as-yet-unanswered questions which are pertinent to the establishment of microbial standards and which must be obtained prior to the establishment of standards. These include, in addition to some of the points to which I have already alluded or presented data thereon:

1. The response of these organisms during storage. Data now available from industry show loss of viable organisms during storage (Wornick, 1963). This we intend to study further through the cooperation of the United Fruit Company.
2. The degree of microbial growth following rehydration and subsequent storage at different temperatures in comparison with similar fresh and frozen products stored under the same conditions.
3. Evaluation of the extensive data now available within the freeze-dehydration industry in order to ascertain "good manufacturing practices." I am happy to be able to report that extensive data from one company on the sampling of many tons of product over a three-year period show microbial counts of very low magnitude on some freeze-dried cooked products, and many lots even approached sterility (Wornick, 1963).
4. The basic philosophical consideration that standards are no substitute for "good manufacturing practice" as a tool for the evaluation of the sanitary quality of foodstuffs or of a food plant itself.
5. Consideration of measurements other than total plate count dependent upon the particular foodstuff and its handling.

It is hoped that those who are empowered with "enabling" legislation will indeed consider the aforesaid prior to any examination of microbial standards. From the experiences of the Association of Food and Drug Officials of the United States (AFDOUS) code, both government and industry have learned much in this respect, to the mutual benefit of both segments and of the consumer.

Conclusion

Freeze dehydration of foods is a relatively new industry but a burgeoning one. The industry so far has taken the lead in utilizing practical microbiological procedures (proved with foodstuffs processed by other means) to assure itself of the microbiological safety of its products. Industry thus not only recognizes its obligations but also is well aware of the stigma and fatalism of one single accident of microbial origin in products produced by a new industry. For this reason industry is to be congratulated on the leadership it has taken in obtaining data on the microbiology of freeze-dehydrated foods.

We in the university must also recognize our role in helping industry and government by seeking out in the laboratory basic scientific truths. Thus, upon completion of our methodology studies for any given foodstuff, we have initiated a program for studies on rehydrated products stored at different temperatures and with different "normal" and abnormal inocula in order to ascertain what happens under "unusual" and "abnormal" conditions. Studies of this type will complement the basic research described earlier and will provide proper, accurate, and meaningful tools for assuring the safety and wholesomeness of this new method of food preservation. To this end, through the generous support of the Public Health Service, our program on the microbiological considerations of freeze-dried foods is dedicated.

References

Davis, B. D., 1950, Studies on nutritionally deficient bacterial mutants isolated by means of penicillin, Experientia 6, 41-50.

Karel, M., 1960, Some effects of water and of oxygen on rates of reactions of food components, Ph. D. Thesis, Massachusetts Institute of Technology.

Levine, M., 1961, Facts and fancies of bacterial indices in standards for water and foods, Food Technol. 15 (11), 29-39.

Silverman, G. J., J. T. R. Nickerson, D. W. Duncan, N. S. Davis, J. S. Schachter, and M. M. Joselow, Microbial analysis of frozen raw and cooked shrimp: I. General results, Food Technol. 15(11), 455-458

Silverman, G. J., N. S. Davis, J. T. R. Nickerson, D. W. Duncan, I. Tezcan, and M. Johnson, 1961b, Microbial analysis of frozen raw and cooked shrimp: II. Certain characteristics of Staphylococcus isolates, Food Technol. 15 (11), 458-465.

Straka, R. P., and J. L. Stokes, 1957, Rapid destruction of bacteria on commonly used diluents and its elimination, J. Appl. Microbiol. 5, 21-25.

Wornick, R. C., 1963, personal communication.

Chapter 8

EVALUATION OF WHOLESOMENESS IN NEWER METHODS OF FOOD PRESERVATION

Leo Friedman

Introduction

Man spends the greatest part of his time on earth working for the next meal. The vital role of food is recognized in the Scriptures, where in the Book of Genesis, before Adam's fall from grace, we read, "And God said: 'Behold, I have given you every herb yielding seed, which is upon the face of all the earth, and every tree, in which is the fruit of a tree yielding seed – to you it shall be for food....' " Later on, after the Flood, God said to Noah and his sons, "Every moving thing that liveth shall be food for you, as the green herb have I given you all."

I am not enough of a scholar to know if elsewhere in the Holy Writ there is any indication that this largesse must be accepted cautiously. I suppose some may interpret the injunction not to eat the forbidden fruit as such a warning, although I believe all would agree that this involves a considerable departure from the usual point of the story. In any event, we know that after countless generations man has learned that relatively few of the multitudinous herbs, seeds, and fruits available to him may be used for food with safety. This lesson must have been learned so painfully that many a new food has undoubtedly been accepted with great reluctance: witness the story of the tomato, which not so long ago was regarded as a poisonous fruit. Experiences with toxic plants have probably contributed to the limited variety of foods, the food customs, and the taboos that exist in different parts of the world.

Many of our established methods of food preservation and food processing such as milling, baking, dehydration, fermentation, brining, and sugaring date back to prehistoric times, and their safety has been tested in the crucible of human experience, not always, however, definitively and with finality. For example, in retrospect, milling of grains with the resulting widespread incidence of beriberi and pellagra could hardly be considered to have been a wholesome practice.

The problem of determining the wholesomeness of food is obviously as old as man, and toxicology is probably one of the oldest areas of human knowledge. Today, no one, king or lord or anyone else, feels the need to have a taster on hand at every meal to test the wholesomeness of every dish served. Mortality and illness due to food poisoning are relatively rare occurrences in most parts of the world. This does not mean that we should be at all complacent about the large number of gastrointestinal disturbances that occur annually because of enterotoxins or otherwise spoiled foods. However, this problem is well understood and, for the most part, can be dealt with effectively on the basis of presently available knowledge. However, in the light of our present knowledge of physiology and nutrition and our concern with metabolic and chronic diseases, the problem of evaluating the wholesomeness of a food or food product is not as simple as it was in the days of the king's taster.

The living organism has often been compared to an enormously complex machine that is in continuous operation, carrying within itself an elaborately equipped maintenance department which makes possible necessary repairs without stopping operation. Furthermore, this extraordinary maintenance department has the ability to enlarge the plant when necessary and desirable and even to make complete working models of the original machine. The obvious requirements for keeping such an unusual machine in continuous operation are: (1) suitable fuel for energy; (2) suitable structural raw materials for maintenance, repair, expansion, or growth; and (3) "operational" materials necessary to maintain proper function which the maintenance department cannot fabricate and which must be supplied preformed.

Nutrition and Food Science

The science of nutrition has been concerned primarily with developing knowledge of the essential nutrients, their nature, their function, and their requirement by men and animals. However, nutrition is much more than the science of the nutrients; it is the science of the metabolism of food and all the substances therein. It must be concerned with their action, interaction, and balance in relation to health and disease, and with all the processes involved in their metabolism. Nutrition is concerned with wholesomeness of food, both from the standpoint of nutritional adequacy for keeping our machine in continuous operation and from the viewpoint of minimizing the effects of those nonnutrient components of food that would tend to gum up the delicate, complex machinery. Food science is concerned with developing knowledge about the chemical, physical, and microbiological reactions and interactions that are needed to make possible the technological developments in agricultural production, preservation, processing, packaging, storage, distribution, and serving of food.

The interests of the nutrition scientist, who is concerned with the metabolism of food and food substances, and of the food scientist, who is concerned with the chemistry, physics, and microbiology of food as it passes from farm to dining room table, converge on the question of wholesomeness. The nutrients and the nutrient value of foods have so far received the lion's share of attention. The food components other than nutrients, whether (1) present originally, (2) added intentionally or incidentally during production or processing, (3) accidental contaminants, or (4) resulting from changes occurring during processing or storage, present very difficult and challenging problems that have been relatively neglected by nutrition and food scientists until recent years.

The potential of promoting food health by dietary means undoubtedly underlies the research activities of many workers in experimental and clinical nutrition. As they explore the possibilities that proper diet may bring about increased resistance to infection, greater vigor, retardation of degenerative disease, and increased useful life span, greater attention will be focused on the nonnutrient components of food. Inevitably, our concepts of evaluating wholesomeness will depart even more from the simple criterion of mortality or serious injury to more sophisticated considerations having a bearing on the health of the individual, even to a small degree.

Food and Drug Laws

The development of the federal food and drug laws represents one of the finest examples of the political and governmental processes whereby the public interest is delineated and defined. It is noteworthy how their very inception and all important changes have been generated by scientific developments and applications. We have witnessed the evolution and acceptance of the philosophy that, in a highly developed and sophisticated society such as ours, the wholesomeness of the food supply can be assured only by procedures operating prior to marketing rather than by punishment of violators after damage has been caused. This philosophy makes good sense for the food industry, since it is a matter of vital self-interest to avoid any situation that could result in a loss of public confidence. The acceptance of an ever-increasing array of new products could not take place if there were any question with regard to wholesomeness. A century of "good will" could easily be nullified by an unfortunate incident. It is important for us all that the American consumer should have complete confidence in the ability of the food industry to furnish, and of the government to assure, only nutritious and wholesome food products in the American market.

Techniques for Safety Evaluation

The responsibility for decisions with respect to the safety of new substances in foods, new processes, and new products is borne by the scientists of the Food and Drug Administration.

During the last thirty years they have established, upon the basis of the best scientific knowledge and techniques available to them, procedures for the appraisal of safety of substances in foods, drugs, and cosmetics. Data obtained by these established and familiar methods may be interpreted with some degree of confidence and are essential for reaching decisions concerning the safe use of new products. Although everyone, especially the FDA people, agrees that these methods are by no means the last word, that better methods capable of yielding more complete and more accurate information more rapidly are needed and should be developed, progress in this direction has been slow. The pressure to reach decisions immediately forces continued use and reliance upon the established procedures. It seems that industry on the one hand, concerned with the profitable development of products, and government on the other, busy with the regulatory aspects of these developments to assure a nutritious and safe food supply, are too much involved with obtaining immediate solutions to daily problems to be capable by themselves alone of maintaining uninterrupted the necessary research and development programs that will result in more effective and rapid methods of safety evaluation.

Evaluation of the safety of a new food product requires, ideally, complete knowledge of the metabolism of all the components and of the physiological effects produced by each, alone and in combination, especially at the levels expected in the ordinary diet. This demands the availability of techniques that will reveal low-level and subtle effects that may be produced by active substances at low concentrations or by relatively inert materials at higher dietary levels. The development of such techniques is an important and challenging scientific problem. A well-rounded and integrated research program is needed that will effectively probe the frontiers of knowledge in all the biological sciences and related disciplines toward this end.

The goals of such a program would include the development of a body of knowledge about the physiological effects of food substances in the various forms they may take in the different circumstances under which they are consumed, including those substances that are known, those that are partly characterized, and those that are only suspected or still unknown. Such a body of knowledge is needed, and it is being developed. It will grow at an exponential rate, and some day will have an appropriate name. At the present time we have settled on the term "food toxicology." No one is altogether happy with this name. Ordinarily a substance comes into the province of toxicology only after incidents of poisoning in man or animal have occurred. The observations of toxic manifestations and mortality serve to stimulate and guide the toxicological laboratory investigations. However, until a better suggestion is forthcoming, "food toxicology" will refer to that scientific discipline concerned with the development of knowledge and understanding of the physiological effects, particularly the subtle effects, of food components that are necessary for proper evaluation of food safety. It must be emphasized that a physiological effect is not necessarily an indication of toxicity.

The presently accepted methodology relies heavily on histopathological examination of tissues of experimental animals that have been subjected to long-term administration of the substance under investigation. In the case of pure compounds or highly purified substances that are relatively potent physiologically, useful positive findings may be obtained. More often, however, decisions must be reached on the basis of negative results. In fact, safety evaluation studies have often been referred to as "the science of negative results." Negative results have their value, but all will agree that positive findings, aside from providing greater satisfaction to the investigator, are much more useful. The use of a "factor of safety," which plays an essential role in applications of data obtained by present methods, is an implicit admission of ignorance — ignorance as to the limits of sensitivity and specificity of our techniques and parameters; ignorance of the degree of comparability of the experimental animal and the human in any particular situation; ignorance as to the significance of observed effects for the health of men and animals. It may not be possible ever to decrease the area of ignorance to the point that will eliminate the need for a factor of safety, but this should be our ultimate goal. Any scientist who has looked scornfully upon "safety testing" would do well to reconsider the magnitude and scope of the scientific challenge presented by the problems just mentioned.

The Research Program in Food Toxicology at M. I. T.

The research program in food toxicology at M. I. T. will be based on the fundamental assumption that every substance is capable of producing a biologic response. The problem we set ourselves is the development of (1) techniques for observing and measuring these effects and (2) a background of knowledge and experience upon which to interpret these observations. We shall approach this problem simultaneously in as many different ways as possible consistent with our resources of people, facilities, and ideas.

Our work will be based firmly upon present techniques. Histopathology, which in many respects is still the most sensitive indicator of biologic effects, will serve as the base line for all our studies. We are unusually fortunate to have as an essential part of our team Dr. Paul Newberne, who is and has been for some time actively engaged in studies of animal nutritional pathology and the relation of nutrition to carcinogenesis. It is our plan to extend the scope of the morphological studies by development and application of electron-microscope and histochemical techniques whenever this promises to be fruitful, particularly in studies of carcinogenesis. We plan also to extend the development and application of other techniques which are at present often used in safety evaluation studies — for example, studies of the metabolic fate of test substances by means of isotopically tagged molecules; tests of physiological function in the whole animal; chemical composition of organs and tissues, including studies of specific enzyme activities.

The last few years have witnessed truly fabulous developments in biochemistry and physiology at the most fundamental levels. These strides in our insights into biological phenomena at molecular and even submolecular levels are occurring at a rapid rate. We should like to eliminate, or at least to shorten, the lag between the development of new knowledge and its application to the problems of understanding the physiological effects of food substances and food processes. I believe it possible and likely that workers with an orientation toward the essentially applied goal of safety evaluation will make significant contributions to our understanding of basic biological phenomena. We should like to have techniques to study, in animals that have been conditioned by a dietary treatment:

1. The functioning of the metabolic machinery — for example, the generation and utilization of energy, the synthesis of various important metabolites or body components, the turnover rate in various compartments.

2. The functioning of various membranes, including intracellular membranes and the endoplasmic reticulum.

3. The functioning of biochemical regulating mechanisms.

Studies along this line will be undertaken with model compounds about which a great deal of knowledge is already available. Since our hope is to observe effects that are beyond the range of our present methods, these studies will be of particular significance for the carcinogenesis problem. We plan to include more than one representative of the several classes of known carcinogens. In all our work we shall include studies both in the whole animal and *in vitro* systems.

From one point of view, the scientific problem may be considered as essentially a problem in bioassay, which is primarily to obtain a living system that is sensitive and will respond consistently and specifically to a given treatment. Toward this end we shall explore the use of different species, strains within species, sex, age, etc., so as to obtain the most susceptible and the most sensitive animal for a specific response. Furthermore, we plan to continue studies of the effect of diet and nutritional status on sensitivity or resistance to specific challenges. For example, newborn or very young rats and mice have been reported to have particular sensitivity in certain carcinogenesis studies. Dr. S. A. Miller of our department has been studying the nutrition of the preweanling infant rat. We plan to apply his experience with this interesting animal in studies of carcinogenesis.

To my knowledge, the possibility that effects observed in older experimental animals may be due to different susceptibility with age rather than to the long-term treatment has not been explored. If this theory can be proved, one of the most compelling justifications for lifetime feeding studies may be weakened, for "normal" animals of any desired age can be made as available as weanling animals are at the present time, and the effect of a dietary treatment in a "standardized old" rat can be determined as quickly as the nutritional requirement for growth of a conventional weanling rat.

Much of the current knowledge and understanding of biochemical and physiological processes has been derived from the study of microorganisms. As a result of intensive work, particularly in virology, there is now available considerable experience in the maintenance of human and mammalian cell lines in tissue culture. If, as some preliminary evidence indicates, human cell cultures reflect the relative susceptibility to toxic agents of man as compared to experimental animals, we may have potentially a most useful tool to bridge the gap between observations in experimental animals and their application to human beings. In vitro cell culture makes possible the study of biological effects by established biochemical techniques in relatively simple systems. We expect that the study of the effects of substances on cell growth, cell morphology, and cell function will provide leads to promising areas for study in intact animals, and conversely indications from animal studies may be followed up more intensively in the simpler system. The use of human cell lines should have some advantage as a supplement to animal experiments in extrapolating with greater confidence than is now possible experimental animal findings to the human situation.

Investigations of this kind, as well as studies with other microorganisms, fertile hens' eggs, etc., as possible test systems, are an integral part of our "food toxicology" program. The laboratory is in the process of being organized and equipped, and studies have been initiated by Dr. Jan Gabliks, a recent addition to our staff.

Besides investigations of new approaches using typical model compounds, we intend to maintain a concurrent program of study on several problems of current interest, using conventional methods, and dovetailing the new techniques when studies have progressed to the point where use of a new approach would be mutually profitable to both investigations.

Research on the "Chick Edema" Factor

One problem of continuing interest concerns the hydropericardium factor, or the so-called "chick edema" factor. A federal regulation requires that fatty acids used in foods be demonstrated by a specified test to be free of this factor. There is still no answer as to the source of this contamination. Until this question is answered, this factor represents a potential threat to the safety of the food supply that should not be forgotten. It has only been found as a minute contaminant, even in the most toxic samples that have been examined. However, its potential hazard derives from its extremely high order of toxicity, not only to the chick, but to rats, mice, guinea pigs, dogs, and monkeys. Most of the observations have been made with crude materials, except in the case of chicks and monkeys. Since this material is so hard to obtain in relatively pure form, very little study has been done thus far.

Research on the Toxic Metabolite Produced by Aspergillus flavus

Another problem that is attracting a great deal of interest centers around the renewed awareness of the potential hazard of the molds to the food supply of man and animals. My colleague and close associate in the food toxicology program, Professor G. N. Wogan, has led the studies at M. I. T. on the production of toxic metabolites by the mold *Aspergillus flavus Link ex Fries*, and their isolation and characterization, which has very recently resulted in elucidation of the chemical structure of aflatoxins B and G, the two most abundant components of the crude toxic extract of food inoculated with this mold.

The progress of this project is a beautiful example of what can be accomplished when people with the necessary knowledge and skill are available and cooperate to advance a project. In addition to Professor Wogan and his immediate group, the contributions of Professor Cecil Dunn on the mycology, Professor Richard Mateles on deep-tank fermentation techniques, Professor Emily Wick on chemistry and isolation procedures, Professors Paul Newberne and William Carlton on the pathological aspects, and Professor George Buchi of the Department of Chemistry have all played essential roles in the progress of the work thus far. Of inestimable value in hastening our progress was the timely cooperation of the British workers who were generous with cultures of the toxin-producing molds, and with descriptions of their most recent knowledge and techniques when the work was first started, and of my former colleagues in the Food and Drug laboratories in Washington, D.C. The latter reciprocated the helpful cooperation they found here when they started their own studies by supplying us with samples of their mold cultures, which provided several times greater yield than any of the cultures we had been using, and with a substantial amount of crude aflatoxin extract. This project will continue to be of major interest as the work on the chemistry of the other components continues and as studies in depth of the toxicology of these substances are undertaken. As efforts to synthesize the aflatoxins continue, we are anticipating studies with isotopically tagged toxins, etc. From such interests, sooner or later a study will develop on the mycology in one or more food industries that will reveal whether and to what extent a mold-related hazard to food safety exists.

Food Fats

Another area of interest to us is that of the fats in our foods, the changes that take place before they get to the dining table, and how these changes influence the wholesomeness of the foods we eat. Specifically, I should mention the fatty acids that do not form urea adducts. These substances are always present in our food, but we should know more accurately to what extent. There is sufficient evidence to indicate that there is no serious short-term hazard. In my opinion, we should have much more information to evaluate properly the long-term hazard, if any. A question of so much public health significance should not go unanswered. It deserves as much attention, at least, as is given to the clearance of any food additive.

Wholesomeness of Irradiation-Preserved Shucked Clams

Since our department has been one of the pioneers in developing radiation processing of food — and a great deal of know-how and experience with marine food products and irradiation techniques is available here — we believe we can make an additional contribution by undertaking the wholesomeness evaluation of low-dose irradiated shucked clams. The United States Atomic Energy Commission would like to have Food and Drug Administration clearance for the application of this process to clams and several other foods. We also see an opportunity to develop data necessary to the acceptance of improved testing procedures.

Problems in the Evaluation of Wholesomeness of a New Food Product vs. That of a Pure Compound

When we evaluate a new food product, a new food process, or the effect of storage conditions or handling on the wholesomeness of a food, we have a somewhat different problem from the evaluation of a pure compound, or a relatively homogeneous stable material that can be tested at many times the level that will be used in food. As has been pointed out earlier, the microbiological and sanitation problems that produce acute effects are fairly well understood. It is more difficult to deal with the problem of small amounts of chemical components producing more subtle effects, which over a longer time may be equally serious. Some of these problems — such as the mycotoxins, heated fats, and accidental contaminants — have been more or less defined. In evaluating the effect of a process such as freeze dehydration, high-temperature short-time sterilization, sterilization by foaming or the use of microbial filters, or chemical preservation by fumigants, antioxidants, antimycotics, antibiotics, etc., or by radiation preservation, there is usually no indication as to where there may be a problem. How, then, shall the investigation of the effect of processing on wholesomeness be conducted?

The first aspect to be considered is the question of nutritive value. All of the nutrients may be measured. Where the absolute values may be in doubt, the relative values are useful to indicate whether a significant alteration in nutrient content has occurred. Biological tests for protein quality and growth studies to detect gross changes in over-all nutritive value are fairly well standardized, of relatively short duration, and not particularly difficult to interpret.

The other aspect of a wholesomeness study is to test whether chemical or other changes attributable to the process have resulted in the presence of new substances in the final product, and if so, whether they are, or may be, present in amounts that would represent a hazard to health.

The accepted procedures for determining the wholesomeness of a food are based on the feeding of the food, in the state in which it is usually consumed in the human dietary, to experimental laboratory animals during a major portion of their life span. The test food must be fed at a level that will make possible a nutritionally balanced diet. There is some question whether such a test, laborious, time-consuming, and expensive as it is, can always provide data upon which decisions may be based with confidence. With positive results, interpretations are relatively easy, but with negative data we are in the position of not being certain. However, the test is essentially a bioassay; since there is usually a dose-response relationship, it seems desirable to increase the dose in order to be in the range of the dose-response curve. This may be achieved by fractionation and concentration of the original material, well-established practices associated with the discovery, isolation, and characterization of most natural substances of biological significance.

Of course, such procedures may have many pitfalls; and until data obtained this way are correlated many times with the established test, they cannot be acceptable substitutes.

Another problem in the testing of many usual foods is their very high moisture content, which presents many difficulties in the preparation and storage of experimental rations and other practical problems of managing an animal experiment. Freeze dehydration has developed to the point where it may be considered as a tool to overcome this problem. Here again, before it can be acceptable as an alternative to the established procedures, this process must itself be thoroughly evaluated and proved not to introduce new factors or artifacts that will confound the interpretation of the results.

Here we have a dilemma. The suggested exploratory procedures cannot be required by the Food and Drug scientists, since alone they will probably not supply acceptable data useful for reaching a decision on wholesomeness. Because the standard techniques are sanctioned by the FDA, no one can see spending the extra effort and money to do the "unnecessary" work. As in the past, the result is that there is little chance for a revision of the "bible" describing procedures for "Appraisal of the Safety of Chemicals in Foods, Drugs and Cosmetics" until the Food and Drug scientists themselves investigate new approaches and techniques of safety evaluation.

In our proposal to the AEC regarding the "Investigation of Wholesomeness of Radiation-Pasteurized Marine Products," we suggest that the development of data showing whether freeze dehydration will significantly alter the food for purposes of wholesomeness investigations is of sufficient import to justify inclusion of a group receiving freeze-dehydrated irradiated clams for comparison with a group receiving the whole irradiated clams. Furthermore, in order to provide a more stringent test for the presence of biochemically active substances in irradiated food, as well as to explore the practical problems that may come up during the long-term feeding program, we propose to carry out ninety-day studies on weanling rats that will be fed extracts of the various test samples, namely: (1) an aqueous extract, (2) a chloroform-methanol extract, (3) concentrated volatile organic matter collected during freeze drying, and (4) the residues from these extractions, in comparison to similar rats fed whole clams and whole clams freeze-dehydrated. This type of study may uncover undesirable effects on wholesomeness of the treated food much earlier than will the conventional feeding of the product as it is ordinarily consumed. With sufficient experience and correlation between this approach and the standard procedure, one may hope that some day the longer study will become obsolete.

Development of methods that are more sensitive, more stringent, and more capable of detecting subtle effects should not be resisted as attempts to prove that everything is harmful, but should be welcomed as being capable of giving us the facts. For, with knowledge of the facts, the food industry will have a sound basis for utilizing its scientific skills and technological resources to produce foods that are more nutritious and more wholesome, foods that are better able to provide the nutritional environment conducive to optimum health.

Conclusion

I have often asked myself, "Is a program of the scope and magnitude indicated realistic?" My answer is, " Not only is it realistic; it is absolutely necessary if we are to have methods and parameters that will give us the answers we must have more conveniently, more completely, more quickly, and more economically."

Such a program has its best chance of success in a department such as ours with its integrated, multidiscipline approach to all aspects of nutrition and food science and in a university where many other specialized scientific skills and talents are available. I am confident that, with the resources of knowledge, skills, interest, and imagination we can muster here in an atmosphere of academic freedom, and with the material and moral support of the Institute and supporting agencies, we shall make significant contributions along the lines we have outlined, and I hope serve to stimulate the development of similar programs elsewhere.

INDEX